EAST-WEST TRADE, INDUSTRIAL CO-OPERATION AND TECHNOLOGY TRANSFER

To Anne, David and Kathryn

East-West Trade, Industrial Co-operation and Technology Transfer

The British Experience

MALCOLM R. HILL
*Loughborough University
of Technology*

Gower

Published by
Gower Publishing Company Limited,
Gower House, Croft Road, Aldershot, Hants GU11 3HR,
England.

 British Library Cataloguing in Publication Data

Hill, Malcolm R.
  East-West trade, industrial cooperation and
  technology transfer.
  1. Great Britain——Commerce——Europe, Eastern
  2. Europe, Eastern——Commerce——Great Britain
  I. Title
  382'.094'047   HF3508.S/

  ISBN 0-566-00591-3

Printed and Bound in Great Britain by
Robert Hartnoll Limited, Bodmin, Cornwall.

Contents

LIST OF TABLES   x

LIST OF FIGURES   xii

FOREWORD by Professor C.H. McMillan, Institute of Soviet & East
          European Studies, Carleton University, Ottawa, Canada. xiii

INTRODUCTION   xiv

ACKNOWLEDGEMENTS   xvi

GLOSSARY   xviii

CHAPTER 1   CHANNELS OF EAST-WEST TRADE, INDUSTRIAL CO-OPERATION
            AND TECHNOLOGY TRANSFER   1

            Introduction   1

            Channels of Trade, Industrial Co-operation and
            Technology Transfer at the Company Level   1

                  Imports and exports   1
                  Inter-firm industrial co-operation and
                    technology transfer   5
                  Technical co-operation agreements   7

            Channels of Trade, Industrial Co-operation and
            Technology Transfer at the Inter-governmental Level   7

                  Inter-governmental trade arrangements and trade
                    agreements   7
                  Inter-governmental agreements of technological
                    co-operation   8
                  Inter-governmental agreements of economic,
                    scientific, industrial and technological
                    co-operation   9

            Notes   11

CHAPTER 2 THE EVOLUTION OF EAST-WEST TRADE, TECHNOLOGY
           TRANSFER AND INDUSTRIAL CO-OPERATION   13

           Introduction   13

           World Trade   14

           The Volume of East-West Trade   15

           The Structure of East-West Trade   20

           Trade in Licences   33

           Industrial Co-operation Agreements   37

           Technical Co-operation Agreements   39

           Trade and Technology Transfer through Inter-
           Governmental Agreements   40

           British Trade Performance   41

           Notes   44

CHAPTER 3 EAST-WEST TRADE AND TECHNOLOGY TRANSFER - THE CASE
           OF SOVIET IMPORTS OF BRITISH MACHINE TOOLS   49

           Introduction   49

           Case Studies of Soviet Imports of British Machine Tools   51

                Case study no. 1 - a producer of special purpose
                   machining systems   51
                Case study no. 2 - a producer of numerically-
                   controlled machine tools   54
                Case study no. 3 - a producer of turning machines   55
                Case study no. 4 - a producer of gearcutting
                   machines   58
                Case study no. 5 - a producer of special-purpose
                   machining systems   60
                Case study no. 6 - a producer of turning machines   63
                Case study no. 7 - a producer of grinding, gear-
                   cutting and turning machines   65
                Case study no. 8 - a producer of automotive
                   components   67

           Comments and Conclusions   69

           Notes   72

CHAPTER 4 EAST-WEST INDUSTRIAL CO-OPERATION AND TECHNOLOGY
           TRANSFER - CASE STUDIES OF BRITISH FIRMS IN THE
           ENGINEERING SECTOR   75

           Introduction   75

Case Studies of British Engineering Companies Engaged
in Industrial Co-operation with Eastern European
Foreign Trade Organisations   78

    Case study no. 1 - a designer and manufacturer of
      mobile cranes   78
    Case study no. 2 - a designer and manufacturer of
      mining equipment   81
    Case study no. 3 - an automotive and aerospace
      engineering company   83
    Case study no. 4 - a designer and manufacturer of
      power transformers   85
    Case study no. 5 - a manufacturer of transport
      equipment and associated components   87
    Case study no. 6 - a designer and manufacturer of
      automotive shock absorbers   89
    Case study no. 7 - a manufacturer of computer
      equipment   90
    Case study no. 8 - a nuclear engineering company   93
    Case study no. 9 - a manufacturer of agricultural
      equipment   94
    Case study no. 10 - an aero-engine manufacturer   95

Comments and Conclusions   97

Notes   99

CHAPTER 5 LARGE SCALE INDUSTRIAL CO-OPERATION AGREEMENTS BETWEEN
BRITISH COMPANIES AND THE SOCIALIST COUNTRIES OF
EASTERN EUROPE   102

Introduction   102

Trade and Industrial Co-operation between British
Aerospace plc and Romania   102

Industrial Co-operation between Massey-Ferguson-
Perkins Ltd and Poland   120

Licensing and Technology Transfer between GKN
Contractors Ltd and Poland   124

Comments and Conclusions   126

Notes   130

CHAPTER 6 BRITISH PURCHASE OF LICENCES FROM THE SOCIALIST
COUNTRIES OF EASTERN EUROPE   133

Introduction   133

The Case Studies   134

    Case study no. 1 - an electronics company   134
    Case study no. 2 - a steel company   135
    Case study no. 3 - the chemical division of a
      steel company   136

Case study no. 4 - a plant engineering company   139
Case study no. 5 - a manufacturer of rechargeable
  batteries   141
Case study no. 6 - an iron foundry   143
Case study no. 7 - a foundry in the aerospace
  industry   144

Comments and Conclusions   146

Notes   147

CHAPTER 7   TECHNICAL CO-OPERATION AGREEMENTS BETWEEN BRITISH
COMPANIES AND THE SOCIALIST COUNTRIES OF EASTERN
EUROPE   148

Introduction   148

The Survey   149

The Case Studies   150

Case study no. 1 - an electrical engineering
  company   150
Case study no. 2 - a photographic equipment
  company   151
Case study no. 3 - the National Coal Board   153
Case study no. 4 - an oil company   155
Case study no. 5 - an electronics company   156

Comments and Conclusions   157

Notes   158

CHAPTER 8   COMMENTS AND CONCLUSIONS, WITH SOME INTERNATIONAL
COMPARISONS   160

Introduction   160

The Performance of British Exporters of Capital Goods
to the Socialist Countries of Eastern Europe   160

Counterpurchase   162

The Role of Project Leaders   162

Industrial Co-operation   163

The Influence of Inter-Governmental Relations   179

General Performance in International Markets   180

Market Preferences of British Companies   181

The Purchase of Licences from the Socialist Countries   187

Tripartite Industrial Co-operation   187

Technical Co-operation Agreements   190

Impact on East European Industries   190

British National-Economic Policies   190

Recent Trends in East-West Political and Economic
Relations   191

Notes   192

APPENDICES   195

APPENDIX A   Questionnaire for Survey of Machine
             Tool Imports   197

APPENDIX B   Questionnaire for Survey of Practices
             in Industrial Co-operation   200

APPENDIX C   Questionnaire for Survey of British
             Purchases of Soviet and East European
             Licences   203

APPENDIX D   Questionnaire for Survey of Technical
             Co-operation Agreements   205

APPENDIX E   A Case Study of West German/Polish
             Industrial Co-operation in Construction
             Machinery   206

APPENDIX F   Market Shares for 'Machinery Non-Electric'   209

BIBLIOGRAPHY   210

INDEX   215

Tables

Table 1.1 Dates of Inter-Governmental Trade Arrangements between the UK and the Socialist Countries of Eastern Europe   7

Table 2.1 World Trade (1958-79)   16

Table 2.2 Trade between the Socialist Countries of Eastern Europe (SCEEs) and the Developed Market Economies (DMEs)   18

Table 2.3 Data for Figure 2.2 - Exports of Engineering Products (1958-78)   21

Table 2.4 Data for Figure 2.3 - Exports of Chemicals (1958-78)   22

Table 2.5 Data for Figure 2.4 - Exports of 'Other Manufactured Goods' (1958-78)   23

Table 2.6 Data for Figures 2.5 and 2.6 - Exports of Food, Crude Materials and Mineral Fuels (1958-78)   24

Table 2.7 Structure of Exports of the Developed Market Economies to the Socialist Countries of Eastern Europe and the World   25

Table 2.8 Structure of Exports of the Socialist Countries of Eastern Europe to the Developed Market Economies and the World   26

Table 2.9 Foreign Trade Data for the USSR and the Remaining Socialist Countries of Eastern Europe (1971-78)   34

Table 2.10 UK Trade with the Socialist Countries of Eastern Europe (1958-79)   42

Table 2.11 Western Exports of Engineering Products to the SCEEs (1963-79)   43

Table 2.12 Western Exports of Engineering Products to the USSR (1963-79)   43

Table 5.1 Western Aircraft Exports to Romania (1968-78)   127

Table 5.2 Total Western Aircraft Exports (1968-78)   128

Table 5.3 Western Machine Tool Exports to Poland (1975-78) 130

Table 8.1 Western Exports of Machine Tools to the USSR (1969-78) 164

Table 8.2 Features of Industrial Co-operation Agreements
           between Socialist Foreign Trade Organisations and
           Fourteen West European Companies   170

Table 8.3 Western Engineering Exports to Western Europe and
           Eastern Europe   183

Table 8.4 Proportions of Western Engineering Exports to
           English-speaking Markets   184

Table 8.5 Imports of Engineering Products by Western Europe
           and English-speaking Markets   184

Table 8.6 Growth Rates for Imports of Engineering Products
           by Western European and English-speaking Markets,
           and the SCEEs.   186

Table 8.7 British Market Shares for Engineering Products   186

Table 8.8 Western Exports of 'Machinery non-electric'   188

Table 8.9 Proportions of Exports of 'Machinery non-electric'
           for selected Western Countries   189

Figures

Figure 2.1   Trade between the Socialist Countries of Eastern
Europe and the Developed Market Economies (1958-79)   19

Figure 2.2   Exports of Engineering Products (1958-78)   28

Figure 2.3   Exports of Chemicals (1958-78)   29

Figure 2.4   Exports of 'Other Manufactured Goods' (1958-78)   30

Figure 2.5   Exports of Food, Beverages and Tobacco (1958-78)   31

Figure 2.6   Exports of Crude Materials, Oils and Fats, and
Mineral Fuels and Related Materials (1958-78)   32

Figure 5.1   Initial Flight of the first TAROM BAC 1-11   114

Figure 5.2   BAC 1-11 Freight Door under construction   115

Figure 5.3   BAC 1-11 Fuselage Half Shells under construction   116

Figure 5.4   Loading of Completed BAC 1-11 Fuselage Half Shells   117

Figure 5.5   BAC 1-11 Build-up   118

Figure 5.6   Loading of BAC 1-11 Fuselage   119

Figure 8.1   Market Shares for 'Machinery non-Electric'   182

Foreword

The early 1970s saw a conjuncture of domestic and international conditions which favoured a rapid expansion of East-West economic relations. One of the more interesting facets of this expansion was the emergence of new co-operative ventures at the level of the firm. These arrangements offered the promise of better integrated East-West production relations as the basis for a more permanent and stable growth in trade. As the international economic and political climate has deteriorated, however, the important question arising has been to what extent the business relationships cemented in the more favourable, earlier period can endure and develop. In this light, a detailed study of British experience is especially timely.

This is Dr. Hill's second monograph in the area of East-West relations. To both he has brought a unique combination of expertise on Soviet industrial organization and international business practice. The present study focuses more specifically on industrial co-operation as the framework for East-West trade and technology transfer. The case study method employed, based on direct contacts with experienced British firms, provides the opportunity for unique insights into the dynamics of East-West co-operative ventures. It is analysis of material such as that presented here on the processes at work in such arrangements that offers the best hope of understanding the inherent vitality and hence durability of the phenomenon.

The author spent several months with us in 1980 as a Commonwealth Visiting Fellow. We were happy at that time to facilitate in any way we could his research in an area of shared interest. We are gratified that some of the work he carried out at Carleton has contributed to the present, useful study.

Ottawa, July 1982   Carl H. McMillan

Introduction

This book is an investigation into the experiences of British companies in the areas of trade, industrial co-operation and technology transfer with those socialist countries of Eastern Europe which are members of the Council for Mutual Economic Assistance (CMEA or COMECON).

Like its predecessor (Hill, 1978), this book has been written to extend the existing publications on the topics of East-West trade and technology transfer, paying particular attention to the requirements of industrial executives and technologists, government policy-makers and administrators, and students of management, economics and politics. The material contained herein should also be of interest to researchers in this field since it complements the investigations of Hayden (1976), Hanson (1981), Paliwoda (1981), Sternheimer (1980) and Yergin (1980).

The book is introduced by two chapters giving a short description of the various types of East-West trade and industrial co-operation agreement (Chapter 1), and a discussion of the volume and structure of East-West commerce during the 1960s and 1970s (Chapter 2). This second chapter also serves to provide a macro-economic background to the company level studies which follow, and also to investigate the changing position of British exports in the share of Western sales to the socialist countries.

The majority of the book consists of case studies of some thirty British companies engaged in various business activities related to trade, industrial co-operation and technology transfer with the socialist countries of Eastern Europe. The case study method has been used extensively because it is the author's view that this approach provides business executives and students of management, with information on the operational opportunities and problems which may confront them in this area of commercial activity. Furthermore, company-based case studies are also useful to acquaint political scientists, economists and government administrators with specific information on business practice, which may be used as contributions to policy discussions and decisions relating to trade with the socialist countries. Most of the studies relate to engineering companies in view of the author's training, experience and interest in that industry; and the important role played by the engineering sector in trade, industrial co-operation and technology transfer with the socialist countries of Eastern Europe. Consequently, much of the material containing details of the technologies that have been transferred between East and West, should be of interest to scientists and engineers.

The case studies provide various perspectives on the East-West trade and technology transfer processes. Chapter 3, for example, provides examples of eight British companies that have recently exported machine tools to the USSR; these cases are then used to evaluate the manner in which the Soviet Union has attempted to absorb Western technology through its foreign trade process. A further two chapters describe the experiences of British companies that have been engaged in industrial co-operation with foreign trade organisations and industrial enterprises in Eastern Europe. The first of these (Chapter 4) deals with a range of ten small, medium-sized and large scale technology transfers to Bulgarian, Czechoslovakian, Hungarian, Polish and Romanian organisations; whilst the second (Chapter 5) deals exclusively with the large-scale industrial co-operation activities of British Aerospace Limited in Romania, and Massey-Ferguson-Perkins Limited and GKN Contractors Limited in Poland. In both of these chapters, particular attention has been paid to the reasons for both sides entering into the particular business arrangement, and whether these objectives were met in practice.

The following two chapters of the book are concerned with facets of the westward, as well as the eastward, flow of technology. Chapter 6 is concerned exclusively with the westward flow of technology through the purchase of licences from the socialist countries, by seven British companies; whilst Chapter 7 is an investigation of the experiences of five British companies that have signed agreements for the exchange of technical information, and joint research and development, with the socialist countries.

The book concludes with a summary of the experiences of British companies in trade, industrial co-operation and technology transfer with the socialist countries of Eastern Europe, using information from the case studies and other published sources. The concluding chapter also includes some comparisons of the experiences of British companies with some of their counterparts in other Western countries, and makes suggestions for further research on this topic.

Acknowledgements

The research described in this book has been financed from a number of sources, and the author wishes to record his thanks for their provision of funding.

The major British source of direct financial support has been the Social Science Research Council, through a research grant to the author to enable the collection of material for Chapters 5 to 7, and part of Chapter 4, during 1981 and 1982. In addition, the author's university department (Department of Management Studies) at Loughborough financially supported the collection of material for part of Chapter 4 during 1979, for publication in the *European Journal of Marketing* (vol. 14, no. 3), in 1980.

The two major sources of overseas financial support have been the Canadian Commonwealth Scholarship and Fellowship Committee of the Association of Universities and Colleges of Canada, and Stanford Research Institute, Menlo Park, California, USA. The Canadian Committee awarded a Commonwealth Visiting Fellowship to the author, during 1980, for research at the Institute of Soviet and East European Studies, Carleton University, Ottawa; and this visit provided the author with the opportunity to consult all of the materials compiled by Professor C.H. McMillan and other researchers associated with the East-West Project at Carleton. Some of these materials were used to compile Table 8.2 and Appendix E, and to locate further organisations to interview for information relating to Chapters 4 to 7. The Stanford Research Institute provided financial support, during 1978, to collect the material presented in Chapter 3, in conjunction with parallel research on chemical technology transfer undertaken by Dr P. Hanson, Reader in the Centre for Russian and East European Studies (CREES), University of Birmingham. A summary of this combined research has been reported previously in Hanson and Hill (1979).

The majority of the information contained in this book has been obtained from industrial and commercial sources: the author consequently wishes to particularly record his acknowledgements to all of the business executives who were willing to be interviewed to provide information on this project, and comment on earlier drafts of the case studies. Special thanks are due to Mr L.J. Rogers, OBE, MA, Consultant in Aviation, Exports and Countertrade, who provided almost all of the material for the British Aerospace/Romanian industrial co-operation case study in Chapter 5, and also commented on previous drafts of the case. In addition, Mr. K. Carter of the International Technology Group of the Department of Trade until his recent retirement, provided much useful preliminary information and comment.

The author also wishes to thank his many academic colleagues who have been willing to discuss this research at various stages of its progress, but clearly any errors in this publication rest with the writer himself. Professor R.W. Davies of the Centre for Russian and East European Studies at the University of Birmingham provided timely initial advice on the scope and structure of this book during the proposal and early develop-ment stages; and Dr P. Hanson has always shown particular interest in the project and given assistance and critical judgement when requested. In addition Professor C.H. McMillan has provided much invaluable assist-ance to the author, especially during his period of research at Carleton during 1980. Various colleagues at Loughborough have commented on parts of earlier drafts, particularly Mr R.H.B. Condie, Dr M. King, Ms J. Lovenduski and Professor J. Sizer.

The author's university department has provided working facilities throughout the period covered by this research; and granted study leave to visit Canada during the autumn of 1980, and to complete the writing of this book during the spring of 1982. The author wishes to record his thanks for these to Professor J. Sizer, Head of Department of Management Studies, and Professor G. Gregory, Acting Head of Department during 1980 to 1982.

The final typescript was prepared for publication by Mrs S. Spencer, and the illustrations were prepared by Mr E.V. Ball.

The photographs used for Figures 5.1 to 5.6 were kindly provided by British Aerospace plc, Weybridge-Bristol Division.

Loughborough, July 1982   Malcolm R. Hill

Glossary

CMEA or COMECON	Council for Mutual Economic Assistance, comprising a membership of the socialist countries of Eastern Europe (SCEEs), Cuba, Mongolia and Vietnam.
COCOM	Co-ordinating Committee of the North Atlantic Treaty Organisation (NATO) which attempts to establish a unified policy regarding exports from NATO members to the countries comprising the Warsaw Pact. All of the SCEEs are members of this Pact.
DMEs	Developed Market Economies (i.e. the developed market economies of Western Europe, including Greece and Yugoslavia, USA, Canada, South Africa, Australia and Japan).
ECE	(United Nations) Economic Commission for Europe.
EEC	European Economic Community.
ECGD	Export Credit Guarantee Department.
FRG	Federal Republic of Germany (Western Germany).
GDR	German Democratic Republic (Eastern Germany).
OECD	Organisation for European Co-operation and Development.
SCEEs	Socialist Countries of Eastern Europe, (i.e. Bulgaria, Czechoslovakia, GDR, Hungary, Poland, Romania, USSR).

The term 'billion' is used to denote 'thousand million'.

1 Channels of East-West trade, industrial co-operation and technology transfer

INTRODUCTION

This chapter is an account of the usual channels of trade and technology transfer between the socialist countries of Eastern Europe and the industrially developed market economies, paying particular attention to the United Kingdom.

The first section of the chapter describes the various types of channels of trade, industrial co-operation and technology transfer at the company level, and the generally supposed reasons for companies entering into these kinds of business arrangements. This section is intended as an introduction to the more specific kind of information contained in the case study chapters occurring later in the book.

This account is followed by a second section which describes the channels for trade, industrial co-operation and technology transfer at the inter-governmental level, paying particular attention to their relevance as a background for company level activities.

CHANNELS OF TRADE, INDUSTRIAL CO-OPERATION AND TECHNOLOGY TRANSFER AT THE COMPANY LEVEL

Imports and Exports

Trade between the socialist countries of Eastern Europe and the industrially developed market economies occurs as a consequence of commercial transactions between Western companies, and authorised foreign trade organisations based in the respective socialist countries. The organisational features of these foreign trade organisations, and the commercial characteristics of other industrial and trading institutions in the socialist countries of Eastern Europe have been already widely discussed elsewhere [1]; this section of the chapter, therefore, is more specifically concerned with the reasons for the various trade relationships between these organisations and Western companies.

Many of the various studies of the economics and politics of foreign trade of the centrally planned economies, carried out by Western specialists [2] during the 1960s and early 1970s, suggested that the

factors affecting the socialist countries' foreign trade policies were
fairly simple. One of the most prolific authors on the topic claimed in
1966 that:

> 'First among the eastern nations, foreign trade is conducted
> primarily to obtain essential imports. Exports are considered
> not as an end in themselves, but as a means to finance the
> necessary imports.' (3)

The same author (4) ascribed this economic behaviour to the effect of a
series of political decisions, the major of which were:

(a) the drive for rapid economic growth, which usually generates import
requirements more naturally and easily than exportable items; and
(b) state trading, causing foreign trade to be conducted primarily to
obtain essential imports, and controlled by the use of import and
export quotas.

Although the above generalisation may have been quite valid for Soviet
foreign trade behaviour, some recent research by Portes, Winter and
Burkett (5) has suggested that each Eastern European country has tended
to follow its own preferred model for foreign trade. Applying
econometric methods to foreign trade data from four selected countries
(Poland, Czechoslovakia, GDR and Hungary), these researchers found that
the foreign trade policy of Poland only had been influenced to an over-
riding extent by its import requirements. Czechoslovakia was found to
primarily attempt to balance its imports and exports; whilst the GDR
was found to determine its imports and exports from price levels, and
constraints on the current balance of payments. The foreign trade
policy of Hungary, on the other hand, was found to have been chiefly
influenced by the potential for its exports.

It is also apparent that as the socialist economies have expanded and
become more complex, their foreign trade decisions have consequently
been influenced by concepts of efficiency (6) and the international
division of labour (7), which are not too far removed from Western
concepts of comparative advantage. When trading with Western countries,
however, it has been difficult for the socialist countries to apply
these newer approaches in a simple manner in view of the inconvertibility
of Eastern currencies, the limited levels of the Eastern countries'
reserves of Western currencies, and the special features of the Eastern
countries' domestic price systems which do not reliably reflect relative
scarcities. These factors have caused any attempt at the application
of comparative advantage theory to be tempered by the necessity of
setting import priorities on certain products in relation to their
effectiveness in planned economic development, the availability of
credit facilities for appropriate purchases, and the possibility of
securing foreign currency through connected export sales.

In addition to these economic and political factors influencing East
European states in their foreign trade policies, it is clear that
technology has also played a major role. (8) Considering the USSR alone,
Amann, Berry and Davies (9) summarised their 1969 study of Soviet
science and industry by the following statement:

> 'rapid expansion of the Soviet economy since the late 1920s

has resulted in industry in which certain sectors are technically very advanced (e.g. space and military R & D, iron and steel technology) but others, notably in the consumer goods industries, are less developed than in the United States and some other major industrial countries.... The Soviet system has proved capable of encouraging and sustaining and, where required, giving massive support to technical innovation in certain selected priority areas. But administrative barriers between research, development and production, and between different government departments, reinforced by the orientation of the Soviet planning mechanism towards maximum production, have created serious obstacles to technical innovation and hindered the achievement of the scientific and technological objectives of the Soviet government.'

Following a study of the technological level of a range of Soviet industries, carried out by a team of researchers almost ten years after the above cited OECD publication, the chief co-ordinator of this study (R.W. Davies) (10) concluded that:

'In most of the technologies we have studied, there is no evidence of a substantial diminution of the technological gap between the USSR and the West in the last 15-20 years, either at the prototype/commercial application stages or in the diffusion of advanced technology.'

Studies carried out by other researchers have also revealed quite severe shortcomings in the design (11) and quality (12) of Soviet industrial articles.

Similarly, in relation to COMECON as a whole, Wilczynski (13) noted in 1974 that:

'(although) the region has reached high technological levels in communication apparatus and installations, food processing equipment, laboratory apparatus, medical equipment, metallurgy, metal-working machinery, ship-building and certain types of machines...the COMECON countries are well behind the West in most branches of industry including those which are usually regarded as the most propulsive carriers of modern technology, viz, those producing automation equipment, electronic apparatus and equipment, motor vehicles, petrochemicals, plastics and synthetic fibres. COMECON is also lagging in the application of such modern technological status symbols as computers, lasers and nuclear power.'

Hence, there would appear to have been a technology gap between the socialist countries of Eastern Europe and the advanced Western nations over a wide range of industries, and this, in turn, appears to have influenced Eastern European nations in their import policy. The case for this view was strongly argued by Sutton (14) and Wasowski (15) in 1973, although it was this author's opinion that a more detailed study of the whole range of technologies, imports and industrial productive capacity was required before the links between COMECON technology and East-West trade could be accurately determined. The recently published report by Zaleski and Wienert (16), which includes a statistical evaluation of technology transfer and a study of the influence of technology transfers on Eastern economies, is a major step in that direction.

It may be concluded from these cited studies, therefore, that the import of products embodying a high degree of technological advancement has been viewed by the socialist countries of Eastern Europe as a means of updating and refurbishing certain industrial sectors, using specific Western comparative technological advantages. Consequently, one chapter of this book (Chapter 3) has been devoted specifically to the use of machinery imports as carriers of technology, using the specific case of the high level of Soviet imports of Western machine tools.

Finally, there are another set of views which consider Soviet import purchases from Western companies to have been influenced by factors other than an improvement of the technological and economic basis of production, or than concepts of comparative advantage. The holders of these views consider that Soviet economic indicators for the sectoral distribution of hard currency have not been very highly developed, and in lieu of reliable economic evaluation of technology purchases, that Soviet buyers have simply attempted to minimise hard currency costs within the context of a shifting set of preferences and priorities ranked as follows (17):

(a) the purchase of types of technology which directly, or indirectly enhanced military capabilities;
(b) a general preference for 'disembodied technology' (i.e. know-how as opposed to products) to minimise hard currency costs for the purchase of hardware, replacing this by Soviet inputs of research and development;
(c) the purchase of complete sets of inter-related technological equipment ('complexes') have been preferred to the purchase of single items or processes;
(d) the possibility of easy duplication of the technology;
(e) the possible use of the technology in export industries.

The present author considers, however, that more detailed research is required across the whole spectrum of Soviet imports before the validity of the above views can be conclusively demonstrated.

Turning now to Western companies' marketing strategies, it is clear that the centrally planned economies have presented themselves as important markets for many firms, particularly those capable of meeting the socialist countries' requirements for the priority items needed for their industrial development. There are several reasons why Western companies choose to export into any market (18) including economies of scale, rapid growth, and market diversification to limit risk and reduce the problems of fluctuating demand; and it is apparent that most of these factors have also held true for exporting to the Eastern European market. (19) Furthermore, for certain companies in industries having a rapidly moving technology, the socialist countries have presented themselves as important markets for specific proven products and processes approaching a mature stage in their life cycle. (20) Indeed, it can be argued that Eastern Europe presented itself as such an attractive market to many Western companies, that pressures were put on various Western governments by those firms, to ease restrictions over non-permitted exports of products to that region. (21)

Many of the Western companies that have operated successfully in the Eastern European market have found, however, that a particular kind of expertise has been needed for continued and profitable business. First and foremost, the capacity to adapt to a more bureaucratic method of doing business has been necessary; this has included a readiness to submit several modified sales quotations with associated negotiations over quite a long period of time; and a meticulous approach to the preparation and fulfilment of contracts. (22) Other aspects of doing successful business in that region have been influenced by specific economic behaviour characteristics of the centrally planned economies, which are considered in the next paragraph.

As has been previously mentioned earlier in this chapter, certain policies carried out by the Soviet and other Eastern European governments have given rise to a growth in import requirements; but the provision of exports to pay for these imports has been hampered in several ways. In the first place, the rapid economic growth which was aimed for in these economies did not always generate a sufficient variety of items for the export market; furthermore, the sellers' markets which emerged in these economies, partly as a result of taut central planning, led to a general lack of concern for the quality of production, including those goods intended for export. (23) In addition to those problems in the products themselves, attention has been frequently drawn to the low level of expertise of certain Eastern European countries in export marketing their products to the advanced Western industrial nations. (24) Consequently, it has frequently been the case that Western companies have had to be prepared to purchase appreciable quantities of Eastern European produced items in order to secure a contract, particularly if their exported product was not an Eastern European import priority. Several cases in this book, particularly in Chapters 4 and 5, have therefore been included to demonstrate the important role played by such countertrade as an element in the marketing mix.

Finally, it has also frequently been necessary for the seller to assist in the arrangement of appropriate credit facilities for the transaction. The high credit rating of the socialist countries of Eastern Europe during most of the 1960s and 1970s, combined with the existence of lines of credit supported by governments at competitive interest rates to promote East-West trade, have led to few problems of export finance for Western companies wishing to export to that area, however.

Inter-firm industrial co-operation and technology transfer

The first generally accepted working definition of industrial co-operation in an East-West context was provided in 1973 by the United Nations Economic Commission for Europe (ECE) namely

' ....the economic relationships and activities arising from:

(a) contracts extending over a number of years between partners belonging to different economic systems which go beyond the straightforward sales or purchase of goods and services to include a set of complementary or reciprocally matching operations (in production, in the development and transfer of technology, in marketing, etc), and

(b) contracts between such partners which have been identified as industrial co-operation contracts by Governments in bilateral or multi-lateral agreements.' (25)

The ECE report cited above also listed several general types of industrial co-operation based on its working definition, namely licensing with payment in resulting products; supply of complete plants or production lines with payment in resultant product; co-production and specialisation; joint ventures; and joint tendering and construction. In addition, Starr (26) also defined sub-contracting and joint research and development as elements of industrial co-operation.

In 1976, following further experience in the actual operation of industrial co-operation contracts, a meeting of experts on industrial co-operation, set up under the auspices of the ECE, put forward the following definition, which was more specific than that advanced in 1973:

'a contractual economic relationship between two or more enterprises of different nationalities, extending over a longer period, whereby a community of interests is established for the purpose of complementary activities relating to the supply of licences and equipment, development of new technologies, the exchange of information on and the use of those technologies, production and marketing with provision for the settlement in kind of whole or part of the obligations arising from co-operation activities.' (27)

For the purposes of the study described in this book, therefore, an industrial co-operation agreement was considered to contain the following elements:

(a) a transfer of technology from the one partner of the agreement to the other, for the design, development and/or manufacture of a particular finished product. The technology transfer may also have included the sale of appropriate licences, manufacturing equipment, or components; and the training of personnel;
(b) the time interval of the industrial co-operation agreement was usually longer than that associated with a one-off sale;
(c) if appropriate, payment for the technology transferred was made partly in items related to the finished product.

It will be seen, consequently, that the study was concerned only with aspects of what had been defined by McMillan (28) as "inter-firm industrial co-operation"; namely industrial co-operation between two parties having the legal competence to carry out commercial activities, usually a British company and a socialist foreign trade organisation.

In view of the fact that inter-firm industrial co-operation has provided a vehicle for the centrally planned economies to obtain technological know-how or manufacturing plant from the advanced Western industrial nations, with a comparatively low outlay of foreign exchange, it has been regarded favourably by socialist foreign trade organisations, particularly during a time of increasing Eastern European indebtedness to Western banks. Since many Western companies wishing to remain in the Eastern European market have embarked on this type of agreement, two chapters of this book are devoted to the experience of British companies in this area of activity. The first of these (Chapter 4) describes the experiences of a range of companies engaged in a broad spectrum of industrial co-operation activities in different Eastern European countries, whilst the second (Chapter 5) concentrates on the activities of British companies engaged in large-scale industrial co-operation projects with Romanian and Polish foreign trade organisations.

In addition Chapter 6 of this book also includes an account of the experiences of British companies that have purchased licences from the socialist countries, to present a comprehensive account of a technology transfer from East to West, as well as the usual transfer from West to East.

Technical co-operation agreements

Many Western companies, including approximately a dozen British companies, have signed agreements with a State Committee for Science and Technology, or industrial ministries and research organisations, in either one or more of the socialist countries of Eastern Europe. These agreements have usually outlined an area of industrial development in which co-operation was to be pursued, providing the company with an appropriate framework for carrying out extensive discussions and promotional activities with Eastern European industrial, scientific and technical personnel. From the Eastern side, they have provided valuable opportunities for contacts with Western specialists in specific areas of technology.

In certain cases, these discussions have developed into commercial arrangements with an appropriate socialist foreign trade organisation; and consequently the present book includes a chapter (Chapter 7) which reviews the experiences of British companies in this area of activity.

CHANNELS OF TRADE, INDUSTRIAL CO-OPERATION AND TECHNOLOGY TRANSFER AT THE INTER-GOVERNMENTAL LEVEL

Inter-governmental trade arrangements and trade agreements (29)

Almost all of the advanced Western industrial nations have signed a series of trade agreements and trade arrangements with the socialist countries of Eastern Europe. In a similar fashion, a series of trade agreements and trade arrangements were also signed between the governments of the UK and the socialist countries of Eastern Europe during the years 1955 to 1972. A summary of the dates of signing these agreements is given in Table 1.1

Table 1.1

Dates of Inter-governmental Trade Agreements between
the UK and the Socialist Countries of Eastern Europe

Socialist Country	Trade Arrangement with the UK	Long-term Trade Agreement with the UK
Bulgaria	1955,1959,1963,1965	1970
Czechoslovakia	1956,1960,1962,1964,1967	1972
GDR	-	-
Hungary	1956,1960,1963,1964,1968	1972

Socialist Country	Trade Arrangement with the UK	Long-term Trade Agreement with the UK
Poland	1957,1960,1963,1964	1971
Romania	1960,1963,1968,1969	1972
USSR		1959,1964,1969

These arrangements and agreements operated basically as bilateral agreements between the governments of the two countries (but occasionally unofficial bodies were the parties as in the case of the UK-GDR arrangement) for periods of time ranging from two to six years. Although the existence of these agreements was never a necessary precondition for trade to take place between British firms and authorised foreign trade organisations of the socialist countries of Eastern Europe, they did provide an agreed framework for British export trade to be developed towards Eastern Europe, since the articles of the agreement usually covered the following topics:

(a) a statement that both of the contracting government parties would endeavour to create conditions conducive to an increase in trade between the two countries;
(b) a listing of areas of particular trading interest;
(c) broad information regarding laws and regulations pertaining to imports and exports between the two contracting parties;
(d) credit, payment and other facilities for the exchange of goods;
(e) provisions to facilitate trade fairs and exhibitions in each other's territory.

It is also important to note that within these agreements there was scope for negotiation over, and alteration of, the tariffs and quantitative restrictions imposed by Britain on imports from the East European countries. Consequently, through the creation of this inter-governmental framework for the development of trade, it can be argued that British firms have been in as strong a position as any located in other Western nations, to carry out East-West trade. It can also be claimed that they were originally in a stronger position with the USSR at least, since the Anglo-Soviet Long Term Trade Agreement of 1959 was claimed to be the first post-war long term agreement to be signed with the USSR by any Western nation. (30)

Inter-governmental agreements of technological co-operation (31)

During the late 1960s the inter-governmental framework for contacts between British and East European organisations was further extended through a series of agreements for co-operation in the fields of applied science and technology in the years listed below:

Bulgaria	1969
Czechoslovakia	1968
GDR	1973
Hungary	1967
Poland	1967
Romania	1967 and 1974
USSR	1968

The topics covered by these agreements usually included the following:

(a) meetings of working groups to examine the possibilities of co-operation in different fields;
(b) visits of experts and technicians for studies, training, consultations and exchange of views in scientific and technological fields;
(c) facilities for study and research, and opportunities to gain experience in industrial research institutions and industrial enterprises;
(d) exchange of industrial knowledge and technology between industrial enterprises in the two countries, including arrangements in the field of licences;
(e) organisation of technical and scientific conferences and seminars of interest to both sides;
(f) facilities for the exchange of technological and scientific information and documentation including books and films.

As in the case of inter-governmental trade agreements, these inter-governmental agreements for technological co-operation were never a pre-condition for technology transfer, particularly of a commercial nature, to take place between British firms and authorised foreign trade organisations of the socialist countries of Eastern Europe; although they did provide a background for British firms to develop arrangements for the exchange of scientific and technical information with appropriate Eastern European organisations.

Inter-governmental agreements of economic, scientific, industrial and technological co-operation

From the end of 1972, the UK Government ceased the signing of any bi-lateral trade agreements with any of the socialist countries of Eastern Europe, in view of the fact that the October 1972 summit conference of the European Economic Community (EEC) decided to follow a common commercial policy towards the countries of Eastern Europe with effect from 1 January 1973. Since Britain became a member of the enlarged Community in 1973 it has adhered to this common policy. Part of this common commercial policy was to confirm that all of the bilateral trade agreements of both old and new member countries would expire at the end of 1974. (32) The only exceptions were to be a few agreements which were due to expire during 1975, including the UK's trade agreements with USSR and Bulgaria which had been signed during 1969 and 1970 respectively. Furthermore, no more bilateral trade agreements could be made by individual member countries after the end of 1972.

The main practical importance of this was that the tariffs and quantitative restrictions imposed on imports from Eastern Europe could no longer be amended by an EEC member-state in a bilateral agreement with an East European state. Such changes had henceforth to be negotiated by the European Commission on behalf of all EEC members. Prolonged and slow-moving negotiations over EEC-Council for Mutual Economic Assistance (CMEA) commercial relations have subsequently been under way between the Commission and the CMEA Secretariat.

Consequently, after 1972, the EEC member states confined themselves to signing a different type of agreement with the socialist countries of Eastern Europe, namely long term agreements of 'economic, scientific, industrial and technological co-operation', and the dates of the UK's signing of such agreements are given below:

Bulgaria	1974
Czechoslovakia	1972 (this Agreement referred to the 1972 Long Term Trade Agreement)
Poland	1973
Romania	1975
USSR	1975

Although trade, inter-firm industrial co-operation, and technical co-operation agreements can be made independently from any inter-governmental co-operation between the UK and an Eastern European state, such an inter-governmental agreement may facilitate the development of commercial transactions. Long term inter-governmental co-operation agreements sometimes give a list of promising areas for future co-operation, and since these agreements are monitored by senior trade officials from both countries meeting annually in the form of a Joint Commission, they provide a platform for discussing the rate of progress in designated potential areas of co-operation. The British side of the policy-making and mechanics of the Joint Commission are the responsibility of the Commercial Relations and Exports Division of the Department of Trade with technical services provided by the International Technology Group of the Department of Trade. Furthermore, reports of the Joint Commission meetings, which usually take place annually in alternate countries, may also list potential areas of co-operation additional to those listed in the long term agreements. Long term programmes for economic, industrial, scientific and technological co-operation can also be drawn up under the auspices of the appropriate agreement, which may list those industrial sectors for which co-operation is most promising. These lists are sometimes divided into

(a) supply of production equipment;
(b) equipment of factories through a turnkey arrangement;
(c) supply of complete plants or integrated complexes of production equipment with payment in the resultant product;
(d) joint production;
(e) research and development and exchange of technical information.

It is important to note, however, that operational arrangements for co-operation agreements are different for the USSR compared with the other Eastern European countries, since in the case of the former a series of Anglo/Soviet Working Groups have been set up to guide co-operation in specific industrial sectors. The British part of the administration of these working groups has been carried out by the Confederation of British Industry (CBI).

NOTES

(1) The organisational and institutional frameworks that have been developed for East-West trade are described in a number of publications, summarised in Hill (1978), pp.50-65.

(2) These studies have recently been reviewed by Turpin (1977), pp.3-14, and summarised in the form of a 'received' Western view of East European, and particularly Soviet, foreign trade policy. These 'received' views are subsequently compared by Turpin with the published statements of Soviet officials (see pp.15-28) and tested against available statistical evidence (see pp.39-52).

(3) See Holzman (1966), p.246.

(4) See Holzman (1966), pp.237-263.

(5) See Portes, Winter and Burkett (1980).

(6) See Brainard, (1976) ('Soviet Foreign Trade Policy') in US Congress Joint Economic Committee (1976), pp.695-708 for a description of the efficiency of these calculations.

(7) See for example, Emgert and Reich (1977) ('Concentration, Specialisation and Co-operation in the CMEA region') in Saunders (1977), pp.79-94.

(8) The notion of comparative advantage can be extended to cover international product flows generated by differences in national technological levels. It is useful, however, to separate technology-based flows from those that can be accounted for by traditional comparative advantage theory relating only to differences in national endowments of capital, labour and national resources.

(9) See Amann, Berry and Davies (1969), pp.487,488.

(10) See Davies (1977), p.66, although Davies did make certain reservations concerning his conclusion, chiefly related to the problems of whether the sample of selected technologies was representative, and that there was also evidence that the position of some technologies was then beginning to change.

(11) See Hutchings (1976), pp.140-240.

(12) See Berliner (1976), pp.301-360.

(13) See Wilczynski (1974), pp.348-353.

(14) Sutton (1973), p.XXV argues that 'by far the most significant factor in the development of the Soviet economy has been its absorption of Western technology' and that one important avenue for importing this technology has been selective purchases of machinery and equipment to be copied.

(15) Wasowski's edited book (1973) also attempted to link Eastern European import policy with the technology gap, particularly pp.43-72 in which R.W. Judy ('The Case of Computer Technology') considers the particular case of computer hardware and software.

(16) See Zaleski and Wienert (1980) especially pp.67-91,139-162,197-240.

(17) See US Congress Office of Technology Assessment (1981), pp.217,218 for an expression of this view.

(18) See for example, Day (1976), pp.27,28.

(19) See Hill (1978), pp.85-135.

(20) See Hayden (1976), Chapter 3 and Hanson (1981), Chapter 2.

(21) See Sutton (1973), pp.200-203 for a discussion of some of the items removed from the COCOM list in the late 1960s and early 1970s to allow Western firms to export equipment to the Volga Automobile Factory (VAZ) at Tol'yatti. (COCOM is the 'Co-ordinating Committee' of NATO, which attempts to establish a unified policy regarding exports from NATO members to Warsaw Pact members, in order that such exports do not enhance the military capacity of the latter (see

Yergin (1980)).

(22) See Hill (1978), pp.85-135,159-166.

(23) See Berliner (1976), pp.301-360 and Holzman (1966), pp.237-263.

(24) See Hill (1974), although there is clear evidence to indicate that East European foreign trade organisations are taking their export marketing tasks far more seriously, particularly for manufactured products, through a higher level of investment in marketing channels in Western countries (see McMillan (1979a) and McMillan (1979b)). Furthermore, some East European foreign trade organisations may argue that their comparatively low level of sales in Western markets are due to discrimination by Western governments and Western buyers.

(25) See ECE (1973).

(26) See Starr (1974), pp.490-492.

(27) See ECE (1976a).

(28) See McMillan (1977a).

(29) The majority of the information for this section is taken from Hill (1978), pp.23-25.

(30) See Macmillan (1971), pp.576 and 615 which mentions discussions prior to the signing of the agreement, and Macmillan (1972), pp.62 and 64 which discusses the signing of the agreement itself.

(31) The information for this section is taken from Hill (1978), pp.23-25.

(32) See Pinder and Pinder (1975), pp.21,22.

2 The evolution of East-West trade, technology transfer and industrial co-operation

INTRODUCTION

This chapter is a review of the growth in volume of trade and technology transfer between the socialist countries of Eastern Europe (SCEEs), and the developed Western economies, to provide a background to the following chapters of the book dealing with commercial arrangements at the company level. The topic has also been covered quite extensively by several economists and political scientists specialising in this area of international economic exchanges (1), and the reader is referred to those sources if a more detailed macro-economic analysis is required.

Throughout this chapter, use has been made of international trade data published by the Statistical Office of the United Nations (2). This source was selected because of its generally accepted credibility, the convenient manner in which trade flow data by product is presented in the Standard International Trade Classification, and the aggregation of trade flows for certain countries into the groups of 'centrally planned economies of Europe and the USSR' (3) and the 'developed market economies' (DMEs) (4).

It is apparent, however, that that published data contains certain inaccuracies, mainly arising from the methods of presentation in the sources of statistics from which the data is compiled. In the first place, many national trade statistics are not submitted to the United Nations in the same classification structure in which the data is to be subsequently aggregated and published, and consequently several assumptions have to be made by the UN Statistical Office, to restructure the source material to the form of the Standard International Trade Classification. Secondly, there may be differences of method used to locate the sources of imports and exports, some countries preferring to classify trade flows by original source and final destination, and others preferring to classify trade flows by most recent source or next destination (5). Finally, there is always the problem of attempting to find a common basis for the comparison of trade flows measured in different national currencies, some of which may fluctuate over the time of collection and aggregation. This latter problem becomes even more acute in an East-West trade context, as a consequence of the basis chosen to convert prices between Eastern European and Western currencies. Nevertheless, in spite of these errors, it was considered by the present author that UN data contained the most accurate and clearly presented information for the purposes of this study.

In the present chapter, it has been decided to select 1958 as a start-
ing point in order to view trade flows until 1978, which was the most
recent data available for foreign trade in certain product groups; although figures for total volumes of exports and imports were also available for 1979. This time interval was selected for two other major
reasons, namely that it was considered to be long enough to show any
trends which might have been emerging; and secondly that it covered the
two decades when most of the trade, technology transfer, and industrial
co-operation agreements were signed.

Estimates of increases have been calculated by comparing volumes of
exports and imports for selected sample years. These were usually 1960,
which pre-dated the increase in East-West trade during that decade; 1972
which marked the commencement of the rapid inflationary increase in
Western prices; and either 1977, 1978 and 1979 as sample years for the
end of that decade. Clearly some criticism can be made against the use
of this method, as there is always the possibility that trade flows for
selected sample years may not be fully representative of the intervening
time periods. Nevertheless, it was considered that this simplified
method was adequate for the present discussion which is concerned with
providing a background to the company level case studies, rather than a
detailed macro-economic analysis of East-West trade. In addition, the
published trade figures are presented in tabular and graphical form in
this chapter to enable the reader to carry out a more detailed analysis
if that is considered necessary.

The problem of accounting for inflation when using foreign trade
statistics in current prices has been avoided to a certain extent by
comparing changes in the value of East-West trade with changes in the
value of trade for the world as a whole. Reference has also been made
to the unit value (price) index, published in the January 1980 Monthly
Bulletin of Statistics for 1960 to 1978, as a measure of the rate of
inflation. Clearly this approach does not fully deal with the question
of inflation, but it was considered satisfactory for the purpose of
comparing growth in trade flows.

The final section of this chapter presents an overview of British performance in the socialist markets, also using available published international statistics. This data, together with some of the information
available from the case studies included later in this book, is subsequently used in the concluding chapter (Chapter 8) to advance views
for the British performance in the capital goods markets of the socialist countries of Eastern Europe.

WORLD TRADE

The foreign trade statistics presented in Table 2.1 illustrate that during the twelve years between 1960 and 1972, world export trade grew
fairly steadily from $128 billion to $413 billion, an increase of some
220 per cent. From 1972, the volume of world trade began to grow more
rapidly and by 1979 had reached some $1,631 billion, or an increase of
almost 300 per cent over that seven year period. Obviously this post-
1972 higher rate of increase was greatly influenced by inflation, as
there was an increase of 130 per cent in the unit value of world exports
from the market economies from 1972 to 1978 when the volume of world
trade increased by some 214 per cent; compared with an increase of only

14

33 per cent in the unit value of world exports from 1960 to 1972 (6), when world exports increased by 220 per cent, as shown above.

The socialist countries of Eastern Europe occupied a comparatively small and slightly declining position in relation to world trade, accounting for some 10 per cent of the receipt and delivery of world exports in 1960, 9 per cent in 1972 and less than 8 per cent in 1979. The developed market economies, on the other hand, accounted for a far larger proportion of world exports, namely 67 per cent in 1960, rising to 72 per cent in 1972 but falling to 66 per cent by 1979. Furthermore, as groups, both the developed market economies and the socialist countries of Eastern Europe have usually been in overall surplus in their total international trading relations, except for 1975 and 1976, when the socialist countries were in slight deficit; and 1979 when the developed market economies were in deficit.

When the figures for world trade with the centrally planned economies are investigated in more detail, however, and the USSR shown separately as in Table 2.1, it becomes apparent that the USSR has been in credit with the world as a whole from 1958 to 1979, except for small deficits in 1972 (less than $400 million in a trade turnover of $31 billion) and 1975 ($2 billion in a trade turnover of almost $69 billion). The remaining socialist countries, on the other hand, as a group had small trading deficits with the world as a whole during the early 1960s, but a larger trade deficit since 1974. In 1978 this deficit reached some $4.5 billion in a total trade turnover of some $125 billion, or approximately 3.5 per cent of total trade turnover.

THE VOLUME OF EAST-WEST TRADE

From the data presented in Table 2.2 and Figure 2.1, it is apparent that East-West trade has increased at a fast rate since 1960: Western exports to the socialist countries increased by some 305 per cent from 1960 to 1972, and by almost the same figure from 1972 to 1979. These increases have been higher than the increases in total world exports calculated in the previous section of this chapter; and also higher than the increase in exports of the developed market economies to the world as a whole (248 per cent from 1960 to 1972, and 262 per cent from 1972 to 1979). On the other hand, the proportion of Western exports delivered to the socialist countries has been low: 2.4 per cent in 1958, 3.1 per cent in 1970 and 3.8 per cent in 1979. Consequently, the socialist countries of Eastern Europe may have presented themselves as growing, but still very small, markets to many Western companies; although some firms witnessed a large growth in orders from the socialist countries during that time period, especially for import-priority items (7).

It is obvious, however, that much of the growth in exports from the Western, to the socialist, countries would have been greatly influenced by inflation in Western prices, particularly since 1972. From 1972 to 1978, the unit value (price) index for the developed market economies grew by some 108 per cent, whilst Western exports to the SCEEs increased by some 238 per cent; compared with an increase in the unit value (price) index of only 31 per cent from 1960 to 1972 (8) when Western exports to the socialist countries increased by over 300 per cent.

The developed market economies have also presented themselves as growth

Table 2.1

World Trade (1958-79)

(All figures in millions of $US FOB)

Year	Total World Exports	World Exports To SCEEs	SCEE Exports To World	World Exports To USSR	USSR Exports To World	World Exports to '6 SCEEs'*	'6 SCEE' Exports to World**	World Exports To DMEs	DME Exports To World
1958	107,880	9,730	10,110	4,140	4,300	5,590	5,810	67,950	71,010
1959	115,370	11,690	11,990	4,930	5,440	6,760	6,550	73,980	75,360
1960	127,870	12,910	12,970	5,360	5,560	7,550	7,410	82,790	85,440
1961	133,970	13,820	14,120	5,610	6,000	8,210	8,120	87,170	90,310
1962	141,410	15,280	15,770	6,290	7,030	8,990	8,740	93,220	94,990
1963	153,860	16,380	17,000	6,840	7,270	9,540	9,730	102,810	103,640
1964	172,160	18,100	18,400	7,590	7,680	10,510	10,720	116,000	117,280
1965	186,390	19,030	19,710	7,810	8,170	11,220	11,540	126,530	128,180
1966	203,400	19,670	20,910	7,640	8,840	12,030	12,070	139,340	141,470
1967	214,190	21,110	22,820	8,280	9,650	12,830	13,170	147,460	149,240
1968	239,140	23,020	24,900	9,110	10,630	13,910	14,270	165,960	167,670
1969	272,710	25,240	27,500	10,050	11,660	15,190	15,840	191,240	193,190
1970	312,348	28,637	30,523	11,451	12,800	17,182	17,723	220,390	224,236
1971	348,266	30,941	33,290	12,108	13,806	18,832	19,484	246,260	250,737

16

1972	413,168	38,272	39,416	15,735	15,361	22,538	24,054	293,554	297,737
1973	517,948	50,390	52,253	20,335	21,462	30,052	30,790	406,079	406,918
1974	838,269	62,694	64,637	24,361	27,405	38,264	37,232	585,780	541,660
1975	872,979	82,665	77,358	35,418	33,310	47,273	44,048	573,197	577,192
1976	989,451	86,977	84,110	36,656	37,169	50,303	46,941	666,433	642,102
1977	1,122,908	95,777	98,106	39,849	45,160	55,880	52,946	749,513	727,709
1978	1,297,518	110,459	112,434	45,847	52,216	64,869	60,218	863,194	871,987
1979	1,631,250	126,477	133,421	53,498	64,762	72,979	68,658	1,116,398	1,079,041

*   i.e. 'World Exports to SCEEs' - 'World Exports to USSR'.
** i.e. 'SCEE Exports to World' - 'USSR Exports to World'.

A footnote to Table B of the *1979 UN Yearbook* notes that this figure does not include trade between the FRG and GDR conducted in accordance with the supplementary protocol to the treaty on the basis of relations between the two countries. The values in millions of $US are as follows:

Year	1974	1975	1976	1977	1978
FRG to GDR	1424	1594	1699	1873	2262
GDR to FRG	865	979	n.a.	n.a.	n.a.

Sources: 1958-1969: *UN Yearbook of International Trade Statistics 1969, Vol. 1*, Table B.
         1970-1979: *UN Yearbook of International Trade Statistics 1979, Vol. 1*, Table B.

Table 2.2

Trade Between the Socialist Countries of Eastern Europe
(SCEEs) and the Developed Market Economies (DMEs) (1958-1979)
(All figures in millions of $US FOB)

Year	DME Exports To SCEEs	SCEE Exports To DMEs	SCEE Surplus With DMEs	DME Exports To USSR	USSR Exports To DMEs	USSR Surplus With DMEs	DME Exports To '6 SCEEs'*	'6 SCEE' Exports To DMEs**	'6 SCEE' Surplus
1958	1,730	1,930	200	610	760	150	1,120	1,170	50
1959	1,940	2,200	260	710	940	230	1,230	1,260	30
1960	2,520	2,520	0	1,010	1,070	60	1,510	1,450	-60
1961	2,710	2,750	40	1,050	1,130	80	1,660	1,620	-40
1962	2,940	2,870	-70	1,270	1,220	-50	1,670	1,650	-20
1963	3,170	3,230	60	1,380	1,360	-20	1,790	1,870	80
1964	3,930	3,660	-270	1,750	1,460	-290	2,180	2,200	20
1965	4,080	4,110	30	1,630	1,650	20	2,450	2,460	10
1966	4,670	4,850	180	1,720	2,000	280	2,950	2,850	-100
1967	5,040	5,240	200	1,870	2,230	360	3,170	3,010	-160
1968	5,380	5,570	190	2,180	2,420	240	3,200	3,150	-50
1969	5,960	6,220	260	2,540	2,580	40	3,420	3,640	220
1970	6,938	7,031	93	2,872	2,716	-156	4,066	4,315	249
1971	7,608	7,761	153	2,809	3,080	271	4,799	4,681	-118
1972	10,199	8,944	-1,255	4,216	3,280	-936	5,983	5,664	-339
1973	15,039	13,392	-1,647	6,196	5,544	-652	8,843	7,848	-995
1974	21,037	19,878	-1,159	8,187	9,156	969	12,850	10,722	-2,128
1975	27,861	20,199	-7,662	13,483	9,582	-3,901	14,378	10,618	-3,760
1976	29,555	23,753	-5,802	14,925	11,607	-3,318	14,630	12,147	-2,483
1977	29,530	26,489	-3,041	14,721	13,436	-1,285	14,809	13,053	-1,756
1978	34,453	29,300	-5,153	16,949	14,356	-2,593	17,519	14,944	-2,575
1979	41,192	39,095	-2,097	20,687	21,373	686	20,565	17,722	-2,883

* i.e. 'DME Exports to SCEEs' – 'DME Exports to USSR'.
i.e. 'SCEE Exports to DMEs' – 'USSR Exports to DME's.

** Also see footnote to Table 2.1 concerning inter-German trade.

Sources: See Table 2.1 above.

18

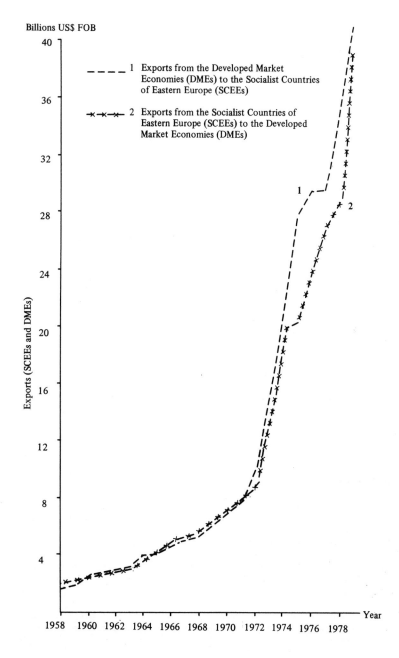

Figure 2.1

Trade between the Socialist Countries of Eastern Europe and the
Developed Market Economies (1958-1979)

Billions US$ FOB

- - - - 1 Exports from the Developed Market
Economies (DMEs) to the Socialist Countries
of Eastern Europe (SCEEs)

―x―x―x― 2 Exports from the Socialist Countries of
Eastern Europe (SCEEs) to the Developed
Market Economies (DMEs)

Exports (SCEEs and DMEs)

Year

1958 1960 1962 1964 1966 1968 1970 1972 1974 1976 1978

markets for products manufactured in the socialist countries, and the westward flow of exports from the centrally planned economies has been increasing at a faster rate (254 per cent from 1960 to 1972, and over 330 per cent from 1972 to 1979) than their growth in exports to the world as a whole (203 per cent from 1960 to 1972, and 240 per cent from 1972 to 1979). Furthermore, the developed market economies have also presented themselves as quite important markets, in total as well as growth terms, for products manufactured in the centrally planned economies. For example, the developed market economies received some 19 per cent of the total world exports of the socialist countries in 1958, some 23 per cent in 1970, and over 29 per cent in 1979.

A final point to consider is the overall balance of trade between the two social and economic systems. Prior to 1970, it was usually the case that the socialist countries had a surplus in their trading relations with the West, commencing at approximately 5 per cent of total trade turnover in 1958 (i.e. 200/3660 x 100 per cent). The post-1970 increase in East-West trade turnover, however, has been associated with a deficit on the part of the socialist countries, accounting for some 16 per cent of total trade turnover in 1975 and 9 per cent in 1978, although the estimate of eastern deficit becomes even higher if CMEA foreign trade statistics are used (9).

When the trade figures are separated for the Soviet Union and the remaining socialist countries of Eastern Europe as shown in Tables 2.1 and 2.2, further information can be obtained. In the first place, it can be seen that since 1976 the Soviet Union, and the grouped remaining socialist countries have each exported and imported approximately the same total values of goods to and from the developed market economies; although for the two decades considered, the Soviet Union has generally traded less than the remaining socialist countries with both the developed market economies and the world as a whole. Secondly, as the value of trade has increased between the two social and economic systems, particularly since 1971, both the Soviet Union and the remaining socialist countries have usually been in deficit with the developed market economies, for their total values of export and import trade; although the USSR returned into credit in 1979. This performance differs from that for their general international trade, since the Soviet Union has usually been in credit in its foreign trading relations with the rest of the world; and the remaining socialist countries have traded at a lower proportion of total deficit compared to their total international trade turnover (see Tables 2.1 and 2.2).

THE STRUCTURE OF EAST-WEST TRADE

This section of the chapter discusses relevant aspects of the structure of East-West trade which, to a certain extent, have influenced the overall values of trade balances discussed above. Considering Western exports to the socialist countries first, data presented in Tables 2.3 to 2.7 shows that machinery and transport equipment, and manufactured goods, have each accounted for some 20 to 35 per cent of Western exports to that region; chemicals and foodstuffs have each accounted for some 10 per cent, whilst mineral fuels have accounted for only 1 per cent. This structure of eastbound exports from the developed market economics, however, is not radically different from the structure of those economies' exports to the world as a whole, since the developed market economies are

Table 2.3

Data for Figure 2.2 - Exports of Engineering Products (1958-78)
(All figures in millions of $US FOB)

Year	DME Exports To SCEEs	DME World Exports	SCEE Exports To DMEs	SCEE World Exports
1958	395	19,790	185	2,820
1959	540	20,680	175	3,500
1960	710	23,820	205	3,730
1961	840	26,150	215	3,740
1962	1,040	28,640	250	4,260
1963	1,000	31,040	250	4,880
1964	1,075	34,960	340	5,560
1965	1,180	39,420	340	6,000
1966	1,490	44,780	415	6,390
1967	1,830	48,980	420	6,880
1968	1,950	56,940	475	7,840
1969	2,260	67,480	540	8,780
1970	2,381	78,620	630	9,600
1971	2,483	91,377	827	10,710
1972	3,309	108,729	947	13,395
1973	4,699	142,307	1,307	17,279
1974	6,333	179,399	1,626	19,636
1975	9,855	212,665	2,191	24,530
1976	10,081	241,884	2,467	26,807
1977	11,012	274,343	2,725	30,906
1978	12,670	326,470	3,131	37,046

Sources: Data for SITC Group 7
1958   *UN Monthly Bulletin of Statistics*; March 1964, Special Table C.
1959-63 *UN Monthly Bulletin of Statistics*; March 1965, Special Table E.
1964   *UN Monthly Bulletin of Statistics*; March 1966, Special Table E.
1965-68 *UN Monthly Bulletin of Statistics*; March 1971, Special Table E
1969-72 *1974 UN Yearbook of International Trade Statistics*; Table B.
1973   *1978 UN Yearbook of International Trade Statistics*; Table B.
1974-78 *1979 UN Yearbook of International Trade Statistics*; Table B.

Table 2.4

Data for Figure 2.3 - Exports of Chemicals (1958-78)
(All figures in millions of $US FOB)

Year	DME Exports To SCEEs	DME World Exports	SCEE Exports To DMEs	SCEE World Exports
1958	120	5,120	145	485
1959	150	5,800	155	540
1960	190	6,490	160	620
1961	205	6,860	185	700
1962	225	7,300	215	790
1963	280	8,090	190	830
1964	365	9,480	185	900
1965	445	10,620	250	1,020
1966	510	11,970	265	1,030
1967	590	12,960	280	1,200
1968	680	14,970	290	1,220
1969	750	17,000	325	1,320
1970	810	19,420	365	1,430
1971	923	21,510	432	1,772
1972	1,126	25,580	488	2,060
1973	1,480	35,610	640	2,530
1974	2,979	56,559	1,214	3,514
1975	3,151	53,205	1,158	4,108
1976	3,073	60,346	1,098	4,097
1977	3,515	68,561	1,291	4,569
1978	4,185	85,312	1,451	5,217

Sources: See Table 2.3 above, but relevant to SITC Group 5 'Chemicals'.

22

Table 2.5

Data for Figure 2.4 - Exports of 'Other Manufactured Goods' (1958-78)
(All figures in millions $US FOB)

Year	DMEs to SCEEs	DMEs to World	SCEEs to DMEs	SCEEs to World
1958	590	21,950	375	2,510
1959	630	23,610	405	2,880
1960	880	27,080	480	3,280
1961	880	27,580	540	3,640
1962	910	28,800	600	4,550
1963	810	31,090	720	4,610
1964	790	36,370	880	5,070
1965	1,030	40,150	1,100	5,410
1966	1,200	44,240	1,310	5,570
1967	1,470	46,160	1,390	6,050
1968	1,620	52,430	1,210	5,990
1969	1,900	61,880	1,410	6,800
1970	2,290	71,040	2,030	7,820
1971	2,627	77,721	2,095	8,321
1972	3,233	91,809	2,443	9,939
1973	4,743	124,190	3,620	12,220
1974	7,893	168,493	4,824	14,670
1975	9,560	168,284	4,219	16,689
1976	9,731	183,910	4,807	17,506
1977	9,345	212,819	6,020	19,967
1978	10,446	259,835	6,854	22,389

Sources: See Table 2.3 above, but relevant to SITC Groups 6 and 8
'Other Manufactured Goods'.

Table 2.6

Data for Figures 2.5 & 2.6 - Exports of Food, Crude Materials
& Mineral Fuels (1958-78)
(All Figures in millions $US FOB)

Year	Food, Beverages & Tobacco		Crude Materials, Oils & Fats		Mineral Fuels & Related Products		
	DMEs to SCEEs	SCEEs to DMEs	DMEs to SCEEs	SCEEs to DMEs	DMEs to SCEEs	SCEEs to DMEs	SCEEs to World
1958	270	470	335	385	3	355	1,340
1959	265	570	335	430	2	395	1,510
1960	330	610	385	600	4	460	1,620
1961	310	730	430	580	2	495	1,820
1962	335	640	400	620	5	540	1,900
1963	610	770	430	630	13	670	2,050
1964	1,410	710	520	790	16	740	2,180
1965	850	810	540	880	15	690	2,260
1966	890	940	540	1,080	12	770	2,300
1967	590	1,040	520	1,150	16	880	2,430
1968	570	940	520	1,180	19	980	2,560
1969	400	1,140	500	1,210	42	1,020	2,710
1970	700	1,120	610	1,240	88	1,170	2,970
1971	880	1,190	579	1,340	74	1,532	3,491
1972	1,580	1,465	806	1,522	90	1,592	3,807
1973	2,605	2,110	1,304	2,410	123	2,790	5,660
1974	1,964	2,187	1,548	3,481	163	5,782	9,402
1975	3,432	2,003	1,477	3,139	206	7,112	13,892
1976	4,483	2,099	1,727	3,121	222	9,017	16,495
1977	3,282	2,124	1,848	3,675	259	10,058	19,555
1978	4,287	2,429	2,243	3,667	349	11,152	22,681

ources: See Table 2.3 above, but relevant to SITC Groups 0 & 1 'Food,
Beverages & Tobacco', SITC Groups 2 & 4 'Crude Materials, Oils & Fats'
and SITC Group 3 'Mineral Fuels and Related Materials'.

Table 2.7

Structure of Exports of the Developed Market Economies to the Socialist Countries of Eastern Europe and the World

Year	Food, Beverages & Tobacco	Crude Materials	Mineral Fuels	Chemicals	Engineering Products	Manufactured Items	Total
Exports to the Socialist Countries (millions $US FOB)							
1958	270	335	3	120	395	590	1,730
1970	700	610	88	810	2,380	2,290	6,940
1978	4,287	2,243	349	4,185	12,670	10,446	34,180
Structure of Exports to the Socialist Countries (%)							
1958	15.6	19.4	0.2	7.0	22.8	34.1	100
1970	10.1	8.8	1.3	11.7	34.3	33.0	100
1978	12.5	6.6	1.0	12.2	37.0	30.6	100
Exports to the World (millions $US FOB)							
1958	10,660	8,500	3,470	5,120	19,790	21,950	70,670
1970	24,270	19,470	7,600	21,850	78,620	71,040	224,180
1978	90,064	56,992	39,877	85,312	326,470	259,835	858,550
Structure of Exports to the World (%)							
1958	15.1	12.0	4.9	7.2	28.0	31.1	100
1970	10.9	8.7	3.4	9.7	35.1	31.7	100
1978	10.5	6.6	4.6	9.9	38.0	30.3	100

Sources: See Tables 2.3 to 2.6 above for volumes of trade.

Table 2.8

Structure of Exports of the Socialist Countries of Eastern Europe
to the Developed Market Economies and the World

Year	Food, Beverages & Tobacco	Crude Materials	Mineral Fuels	Chemicals	Engineering Products	Manufactured Items	Total
Exports to the Developed Market Economies (millions $US FOB)							
1958	470	385	355	145	185	375	1,930
1970	1,120	1,240	1,170	325	630	2,030	7,231
1978	2,429	3,667	11,152	1,451	3,131	6,854	28,684
Structure of Exports to the Developed Market Economics (%)							
1958	24.4	19.9	18.4	7.5	9.6	19.4	100
1970	15.5	17.1	16.2	4.5	8.7	28.1	100
1978	8.4	12.8	38.9	5.1	10.9	23.9	100
Exports to the World (millions $US FOB)							
1958	1,310	1,490	1,340	485	2,820	2,510	10,110
1970	3,150	3,040	2,970	1,430	9,600	7,820	30,523
1978	7,457	8,068	22,681	5,217	37,046	22,389	102,858
Structure of Exports to the World (%)							
1958	13.0	14.7	13.3	4.8	27.9	24.8	100
1970	10.3	10.0	9.7	4.7	31.5	25.6	100
1978	7.2	7.8	22.0	5.1	36.0	21.0	100

Sources: See Tables 2.3 to 2.6 above for volumes of trade.

specialised international traders in chemicals, engineering products and other manufactured goods (see Table 2.7 above). From the data given in Tables 2.3 to 2.5, and presented in Figures 2.2 to 2.4, it can also be seen that during most of the 1970s the purchases of these goods by the centrally planned economies from the developed market economies, generally increased at a rate approximately equal to or higher than total Western exports of these products; although the increase in value of these sales from the mid-1970s would also have been influenced by inflationary tendencies in world market prices as well as real increases in the volume of trade. On the other hand, although the growth in value of sales of chemicals and engineering products from the developed market economies to the socialist countries has been considerable, the Eastern European market has still only accounted for some 4 to 5 per cent of total Western exports of these products during most years.

Turning now to exports from the socialist countries, it is apparent from Table 2.8 that the structure of exports from those countries to the West has been different from their exports to the world as a whole; since the developed market economies have purchased a smaller proportion of engineering products from the socialist countries, but a far larger proportion of food, beverages and tobacco; crude materials; and mineral fuels. The proportions of sales of chemicals and other manufactured items were approximately the same to the developed market economies as to the world as a whole.

This structure of trade has had very important consequences for the overall pattern of trade between the two social and economic systems, since the following is apparent (see Tables 2.3 to 2.8 and Figures 2.2 to 2.6):

(a) prior to 1974 it was usual for trade to be evenly balanced in food-stuffs, beverages and tobacco, with the socialist countries overall usually enjoying a small surplus, although in some specific years they were in deficit. Since 1974, however, this deficit has increased rapidly, mainly due to Soviet purchases of these products;

(b) the socialist countries have incurred a growing deficit in chemicals and other manufactured goods, but particularly in engineering goods. The centrally planned economies have increased their sales of each of these products to the West at a faster rate than to the world as a whole, but their volume of sales has still been far too small to close the trade gap with the developed market economies for these products;

(c) the socialist countries have enjoyed a surplus in their sales to the West of crude materials, oils and fats; and particularly of mineral oils and related materials. The increases in sales of these products to the West have closely followed their increases in sales to the world as a whole, but they have still been insufficient to close the overall trade gap with the developed market economies.

In almost every case considered above, it is also important to note that the Western economies are quite important purchasers of socialist goods, accounting for some 30 per cent of total socialist export sales of food, beverages and tobacco, chemicals, and other manufactured goods; and some 45 to 50 per cent of total socialist exports of crude materials and mineral fuels. The West would appear to be a less important purchaser of engineering goods, however, accounting for only some 10 per

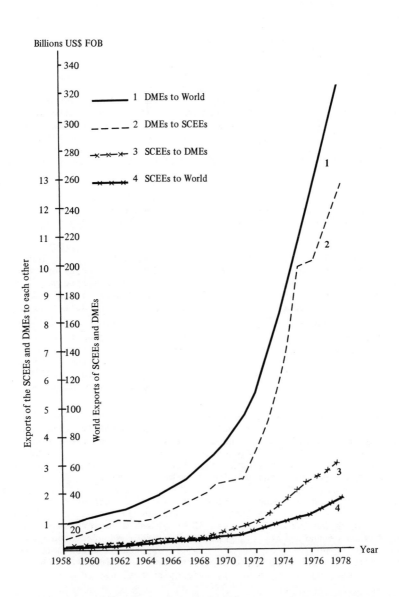

Figure 2.2

Exports of Engineering Products (1958-1978)

Figure 2.3

Exports of Chemicals (1958-1978)

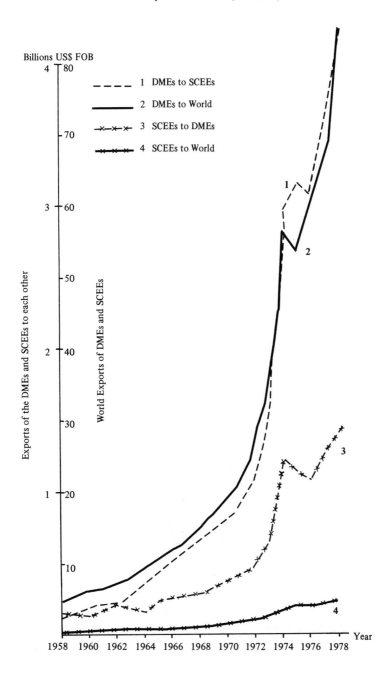

Figure 2.4

Exports of "Other Manufactured Goods" (1958-1978)

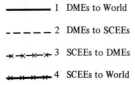

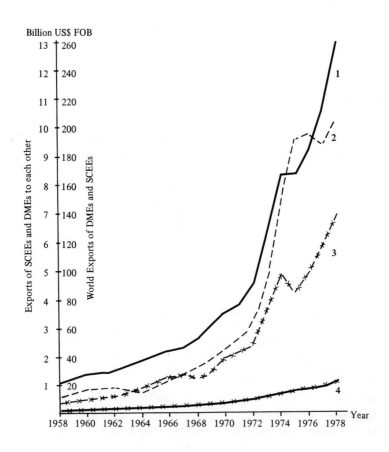

Figure 2.5

Exports of Food, Beverages & Tobacco (1958-1978)

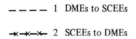

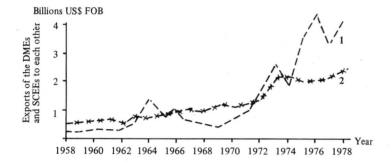

Figure 2.6

Exports of Crude Materials, Oils & Fats; and Mineral Fuels and
Related Materials (1958-1978)

Exports of Crude Materials, Oils & Fats

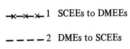

—×—×—× 1   SCEEs to DMEEs

— — — — 2   DMEs to SCEEs

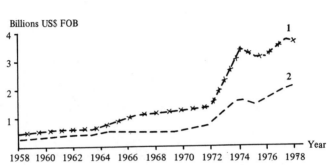

Exports of Mineral Fuels and Related Materials

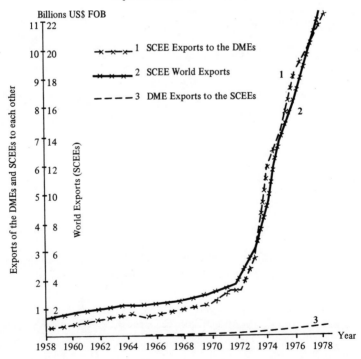

cent of total exports of these products from the socialist countries.

As in the previous section of this chapter, further information can be obtained if the trade figures for the USSR alone are separated from those of the remaining socialist countries of Eastern Europe; although this data for the Soviet Union has only been published in the UN Yearbook since 1971.

It can be seen from Table 2.9 that both the USSR and the remaining socialist countries have incurred quite substantial overall levels of deficit with the developed market economies, for trade in engineering products and chemicals. The size of deficit for chemicals has generally been lower for the Soviet Union than for the remaining socialist countries; but for engineering products the Soviet deficit has been substantially larger during the more recent years.

The USSR has also had a comparatively large level of deficit with the developed market economies for food, beverages and tobacco, and other manufactured products; whereas the remaining socialist countries have been in credit for the former group of products, and have incurred a far lower level of debt for the latter. On the other hand, both the USSR and the remaining socialist countries have been in credit with the developed market economies for trade in crude materials, oils and fats; and mineral fuels and related products. The USSR has had a far larger level of credit for these items than the remaining socialist countries, thereby enabling it to substantially reduce its overall level of deficit with the West.

In conclusion, therefore, the pattern of East-West trade during the years discussed in this chapter, can be considered as a predominantly eastward flow of goods having a high content of embodied technology (i.e. engineering goods, chemicals and, to a lesser extent, manufactured goods), and a predominantly westward flow of raw materials and semi-manufactured goods; although the westward flow of manufactured goods has also increased in value. This pattern of foreign trade between the two social and economic systems, in its turn, is considered by some authors (10) to have indicated the existence of a considerable technological gap between the developed market economies and the socialist countries of Eastern Europe. The remaining chapters of this book, therefore, investigate these aspects of trade and technology transfer in more detail.

TRADE IN LICENCES

It is difficult to obtain accurate data on East-West trade in licences and associated payments because of the lack of detail on this topic in the published statistics, and also because the sale of a piece of processing plant may also include a licence which is not usually itemised as a separate payment. The estimates that have been made on this topic, however, suggest a predominantly eastward flow of licences and associated technological know-how, and a predominantly westward flow of payments, either in hard currency or in goods produced by the licence. This latter form of payment frequently forms part of an inter-firm industrial co-operation agreement, however, and will consequently be discussed in the next section of this chapter which refers to that topic.

Wilczynski (11) estimated in 1976 that the total number of licences

Table 2.9

Foreign Trade Data for the USSR and the Remaining Socialist
Countries of Eastern Europe (1971-78)
(All figures in millions of $US FOB)

Year	DME Exports To USSR	USSR Exports To DMEs	Balance (USSR)	DME Exports To Remaining Socialist Countries ('6 SCEEs')	Exports of Remaining Socialist Countries ('6 SCEEs') to DMEs	Balance (6 SCEEs)
Engineering Products						
1971	983	154	-829	1,500	673	-827
1972	1,283	181	-1,102	2,026	766	-1,260
1973	1,831	273	-1,558	2,868	1,034	-1,834
1974	2,454	356	-2,098	3,879	1,270	-2,069
1975	4,840	547	-4,293	5,015	1,644	-3,371
1976	5,310	615	-4,695	4,771	1,852	-2,919
1977	5,828	652	-5,176	5,184	2,073	-3,111
1978	6,630	752	-5,878	6,040	2,379	-3,661
Chemicals						
1971	315	88	-227	608	344	-265
1972	349	73	-276	777	415	-362
1973	372	112	-260	1,108	528	-580
1974	868	301	-567	2,111	913	-1,198
1975	1,106	323	-783	2,145	833	-1,212
1976	1,018	193	-825	2,055	903	-1,152
1977	1,302	273	-1,029	2,213	1,018	-1,195
1978	1,552	338	-1,214	2,633	1,113	-1,520
Other Manufactured Goods						
1971	1,138	486	-652	1,489	1,609	120
1972	1,418	543	-875	1,815	1,900	85
1973	1,984	1,000	-984	2,759	2,620	-139
1974	3,525	1,227	-2,298	4,368	3,597	-771
1975	4,835	923	-3,912	4,725	3,296	-1,429
1976	5,142	787	-4,355	4,589	4,020	-569
1977	4,874	1,361	-3,513	4,771	4,659	-112
1978	5,412	1,458	-3,954	5,034	5,396	362
Food, Beverages and Tobacco						
1971	287	170	-117	593	1,020	427
1972	892	99	-793	688	1,366	678
1973	1,525	131	-1,394	1,080	1,979	899
1974	800	165	-635	1,164	2,022	858
1975	2,133	156	-1,977	1,299	1,847	548
1976	2,679	111	-2,568	1,804	1,988	184
1977	1,902	192	-1,710	1,380	1,932	552
1978	2,262	202	-2,060	2,025	2,227	202
Crude Materials, Oils and Fats						
1971	165	808	643	414	532	118
1972	250	905	655	556	617	61
1973	445	1,525	1,080	859	885	26
1974	460	2,240	1,780	1,088	1,241	153
1975	456	2,110	1,654	1,021	1,029	8

Table 2.9 (continued)

Year	DME Exports To USSR	USSR Exports To DMEs	Balance (USSR)	DME Exports To Remaining Socialist Countries ('6 SCEEs')	Exports of Remaining Socialist Countries ('6 SCEEs') to DMEs	Balance (6 SCEEs)
1976	630	2,004	1,374	1,097	1,117	80
1977	644	2,452	1,808	1,204	1,223	19
1978	876	2,302	1,426	1,367	1,365	-2
	Mineral Fuels and Related Products					
1971	7	1,054	1,047	67	478	411
1972	7	1,044	1,037	83	548	465
1973	10	2,018	2,008	113	772	659
1974	23	4,195	4,172	140	1,587	1,447
1975	39	5,179	5,140	167	1,933	1,766
1976	40	6,812	6,772	182	2,205	2,023
1977	54	7,981	7,927	205	2,077	1,872
1978	90	8,796	8,706	259	2,356	2,097

Sources: For USSR: Data for 1971 &1972 abstracted from *1976 UN
Yearbook of International Trade Statistics*, Table B.
            Data for 1973 to 1977 abstracted from *1978 UN
Yearbook of International Trade Statistics*, Table B.
            Data for 1978 abstracted from *1979 UN Yearbook of
International Trade Statistics*, Table B.
            For the 'Remaining Socialist Countries' ('6 SCEEs'),
data for the USSR has been subtracted from data for 'Centrally Planned
Economies of Europe and the USSR' shown in Tables 2.3 to 2.7 above.

sold by Western companies to organisations in the socialist countries of
Eastern Europe, had exceeded 2400 (including 500 to Yugoslavia), of which
more had been sold during 1970 to 1975 than during the entire period
prior to 1970; and the entire proceeds to the West from the sale of these
licences were estimated to be some $300 million, or some $240 million if
Yugoslavian royalties are excluded on a pro-rata basis (i.e. $300 million
x 1900/2400). The main purchasers of these licences were apparently USSR
and Poland (about 450 licences each) followed by Czechoslovakia and
Hungary (250 to 300 licences each) (12).

The socialist countries, on the other hand, were estimated to have sold
approximately 700 licences to the West 'mostly involving minor specialised (but not necessarily of low sophistication) inventions, not in the
same class as those purchased from the West' (13). Czechoslovakia was
claimed to be the largest single exporter of socialist licences to the
West (approximately 350 licences), followed by the USSR, Hungary, GDR
and Poland (14). The receipts per licence from the capitalist countries
have been estimated to have been only one-eleventh of the socialist
countries' licence payments to the West during the 1960s (15), but by
the early 1970s it was claimed that this ratio had increased to one-
quarter (16), although the sale of socialist licences to the West still
only amounted to $40 million by the mid 1970s (17). There is also
evidence to show that one socialist country at least, namely Hungary,
had a reduced deficit in licence payments with the West in 1973,
compared with 1970, 1971 and 1972 (18).

To the present author's knowledge, no detailed study has been carried
out on the structure of East-West trade in licences, although Kiser has
provided useful data on the westward flow of licences from the socialist
countries, in two reports published in 1977 and 1980 respectively (19).
The first of these referred to a sample of 116 Soviet licences drawn
from an estimated population of some 300 Soviet licences sold since 1972,
whilst the second referred to approximately 300 licences claimed to have
been sold to Western countries since the mid 1960s by Czechoslovakia, GDR,
Hungary and Poland. The topics to which this data related could be
classified into metallurgy (including process engineering); chemicals and
chemical equipment; medicine and biochemicals; electronics, instruments
and measuring devices; machines and equipment; energy; transportation;
environment (including building and construction); and mining. The
results for the latter four topics in both the Soviet and East European
data were similar, with each topic accounting for less than 4% of the
total in each sample, and there were also similarities for electronics,
instruments and measuring devices - 10% of the total in the case of the
Soviet sample, and 8% of the total for the other East European countries.
There were differences, however, between the two sets of countries for
the remaining topics. The USSR appeared to have sold a far higher proportion of licences for metallurgy than the other East European countries
- 49% compared with 20%; but a far lower proportion of licences for
chemicals - 11% compared with 23%, medicine and biochemicals - 9%
compared with 16%, and machines and equipment - 12% compared with 27%.

No similar survey has been carried out on the eastward flow of licences,
although this structure may be very similar to that of 'licences with
payment in resultant product'. Data (20) available on this topic, for
1978, reveals the following structure: chemical industry - 16%, transport
equipment - 26%, mechanical engineering and electronics - 21% each,
electrical engineering, food and agriculture and light industry - 5% each.

INDUSTRIAL CO-OPERATION AGREEMENTS

The quantity of industrial co-operation agreements between Western companies and foreign trade organisations in the socialist countries of
Eastern Europe, has grown substantially since the late 1960s. A Soviet
source (21) claims that there were some 180 agreements in 1968, which
had approximately doubled to between 350 and 400 by the beginning of
1970. The same source also claims that the quantity of co-operation
agreements subsequently increased to 600 by the end of 1970, although
information in a Hungarian source (22) claims that this volume had not
been reached until 1975.

Using several Eastern and Western sources on industrial co-operation,
Levcik and Stankovsky (23) estimated the number of East-West co-operative
undertakings to have been 600 at the end of 1973, and to have passed the
1,000 mark by 1975. This latter estimate is also in broad agreement
with another report of the ECE (24) which in 1976 quoted the signing of
over 1,000 industrial co-operation agreements, and a similar figure was
also quoted by Bykov in 1977. Further support for this broad estimate
was also provided by the results of a survey to estimate the 'universe'
of industrial co-operation agreements, completed in 1976 at Carleton
University, Ottawa. This survey recorded the existence of 1,076 East-
West 'inter-firm co-operation agreements' and 1,415 co-operation agreements if 'technical co-operation' agreements were included (25).

In conclusion, therefore, it appears that the quantity of inter-firm
industrial co-operation agreements increased by more than four times
during the period from the mid 1960s to the mid 1970s although such an
estimate is tenuous to a very great extent, since all Western compilers
of the quantity of industrial co-operation agreements sound a note of
caution over the accuracy of their estimates. Furthermore it is
difficult to ascertain the degree to which these estimates are independent, and the extent to which the figure of '1,000 industrial co-operation
agreements' was selected as a convenient rounded figure for publicity
purposes. Although the rate of growth of the signing of industrial co-
operation agreements was probably quite rapid over the decade considered,
it is this author's opinion that the total quantity of these agreements
are probably quite small when compared with the total number of East-
West business arrangements.

It is still more difficult, however, to obtain reliable estimates of
the total financial value of co-operation agreements. A publication (26)
by Wilczynski in 1976, estimated such trading activities to have been
worth some $1 billion (presumably in 1974 and 1975), although this
figure may have included some co-operation activities between Western
companies and Yugoslavian organisations, with which we are not concerned
in this present study, since Yugoslavia is not a member of COMECON.
Comparing this estimate with total foreign trade between the socialist
countries of Eastern Europe and the developed market economies, which
was some $43 billion in 1976 according to the 1979 United Nations Year-
book of International Trade Statistics, it appears that industrial co-
operation activity accounted for approximately 2 per cent of total East-
West trade. Other data available for Hungary, a pioneer in the field of
East-West industrial co-operation having signed 400 agreements with
Western European companies by 1975, suggests that the share of turnover
under co-operation agreements amounted to some 2 per cent to 3 per cent
of Hungarian foreign trade with capitalist countries (27), with some 3

per cent to 4 per cent of Hungarian exports to capitalist countries
being accounted for by items manufactured under such agreements (28).
This proportion appeared to increase for certain countries (e.g. 7 per
cent for exports to the Federal Republic of Germany) (29) and also for
certain types of products: it is claimed for example, that approximately
one-fifth of Hungarian exports of machinery to capitalist countries were
manufactured under co-operation agreements (30). A similar pattern was
also apparent for Poland, for which 1.5 per cent to 2 per cent of exports
to capitalist countries were claimed to have been manufactured under co-
operation agreements in 1974, this figure increasing to some 10 per cent
for machinery and equipment (31). Data quoted from Olszynski (32) by
Levcik and Stankovsky (33) also suggests that in 1970, 14 per cent of
Polish exports of capital goods to Western Europe (excluding Finland),
were delivered under co-operation agreements, and in the case of exports
to the Federal Republic of Germany, the proportion was claimed to be as
high as 71 per cent. In conclusion, therefore, it seems that only a
comparatively small proportion of total East-West trade has been manufactured under appropriate co-operation agreements, probably an annual
proportion of less than 10 per cent (34), although the proportion has
probably been higher for the westbound exports of capital goods. As in
the case of estimates of the quantity of industrial co-operation agreements, however, we are faced with the tenuousness of estimates caused by
the fact that certain of the estimates may not be independent, and also
that some of the quoted figures may have been selected for publicity
purposes alone. Furthermore, we are also faced with the same problem
discussed in the section on licensing, namely the degree to which trade
through industrial co-operation may not be reported separately from a
larger trade contract of which it forms a part. Finally, a measure based
on trade turnover alone may not be the best indicator of industrial co-
operation activity, as a small number of large contracts may have
accounted for a comparatively large proportion of trade turnover, although the larger degree of inter-organisational involvement which industrial co-operation was supposed to foster may have been comparatively
small (35).

Turning to the structure of industrial co-operation agreements, the
most comprehensive published research on this topic has been carried out
by the United Nations Economic Commission for Europe (ECE), commencing
with an analysis of some 202 co-operation contracts in 1973 (36).
Similar analyses were subsequently carried out on samples of 297 contracts in 1975, 298 contracts in 1976 and 314 contracts in 1978 (37).

Taking the socialist countries first, it was found that organisations
from Hungary, Poland, Romania and the USSR had concluded approximately
90 per cent of the co-operation agreements between organisations from the
planned economy and the market economy countries. In 1976, Hungary had
concluded 30 per cent of the sample of agreements, Poland 26 per cent,
Romania 20 per cent and the USSR had concluded 13 per cent. By 1978,
the ranking had changed quite considerably with the USSR found to conclude 41 per cent of the sample, Hungary 24 per cent, Poland 17 per cent
and Romania 9 per cent. The ranking of the Soviet Union should be
treated with a certain amount of caution, however, since the ECE may have
included 'technical co-operation' agreements in its sample, and the USSR
has tended to sign a comparatively large number of these types of agreement - 275 of 339 reported 'technical co-operation' agreements recorded
by a Carleton study (38). Another study of a sample of 218 inter-firm
co-operation agreements carried out at Carleton in 1975, and also

reported by McMillan in the previously cited source, showed Hungary and
Poland to be the established leaders in inter-firm East-West industrial
co-operation.

The 1976 ECE report (39) also listed the countries of origin of the
major Western companies that had concluded co-operation agreements with
organisations located in the socialist countries of Eastern Europe, and
these were as follows: Federal Republic of Germany (25 per cent of
sample), France (16 per cent), Austria (12 per cent), USA (9 per cent),
Japan (9 per cent), Italy (8 per cent), Sweden (7 per cent), and the
United Kingdom in eighth place at 6 per cent of the sample. Information
available from the 1976 and 1978 ECE reports (40) also listed the
following information on industries and types of co-operation for those
years:-

(a) the mechanical engineering (including machine tools), chemical,
    transport equipment, electrical and electronic, and metallurgical
    industries accounted for the bulk of industries covered by the
    sample of co-operation agreements (36 per cent, 18 per cent, 14 per
    cent, 11 per cent and 7 per cent respectively in 1976; 22 per cent,
    26 per cent, 10 per cent, 18 per cent and 8 per cent in 1978). The
    remaining 15 per cent of the sample was made up of a wide range of
    consumer industries (textiles, footwear, rubber, glass and furniture)
    and construction, hotel management and tourism;

(b) the major types of co-operation agreement encountered in the sample
    were as follows:-
      co-production based on the specialisation of partners (30
      per cent in 1976 and 45 per cent in 1978);
      the supply of licences, with payments in resultant products
      manufactured under the licences supplied (25 per cent in
      1976 and 6 per cent in 1978);
      the supply of plant and equipment with payment in resultant
      products manufactured by the plant supplied (29 per cent in
      1976 and 17 per cent in 1978).

The ECE reports cited above did also note, however, that there was an
affinity for certain industries to follow certain modes of co-operation.
Co-production, for example, tended to be common in the product-based
engineering industries; the supply of plant and equipment with payment
in resultant product was common in the process-based chemical and
metallurgical industries, and licensing with payment in resultant product
tended to be common in both the engineering and the chemical industries.
The Carleton study also revealed similar patterns, with more than 45 per
cent of the agreements including licensing, technical assistance,
training of Eastern European personnel and production specialisation.
McMillan (41) consequently concluded that the essence of East-West
industrial co-operation lay in 'co-ordinated specialisation in the production of components of an end-product on the basis of transferred
technology'. Using the evidence on licensing payments discussed in the
previous section of this chapter, it is likely that the majority of such
transferred technology has been in an eastward direction.

TECHNICAL CO-OPERATION AGREEMENTS

As explained in Chapter 1 above, 'technical co-operation agreements' or 'framework agreements' are the terms used in this book to describe agreements established for the exchange of scientific and technical information, between Western companies and Eastern European state committees responsible for the overall development and co-ordination of science and technology; and also between Western companies and Eastern European industrial ministries responsible for the development of specific technologies. It has been estimated by McMillan (42) that the number of such agreements had reached some 339 by 1977, 275 of which had been signed by the USSR.

To the present author's knowledge, there has been no detailed survey of the industrial groups into which these agreements can be categorised, although data available from one survey (43) of framework agreements signed by Soviet organisations (sample size of 188 agreements) suggested that they could be classified as follows:-

Chemicals	29%
Machinery	50%
Transport Equipment	7%
Electrical Engineering and Electronics	18%

TRADE AND TECHNOLOGY TRANSFER THROUGH INTER-GOVERNMENTAL AGREEMENTS

As explained in Chapter 1 above almost all of the governments of the socialist countries of Eastern Europe have signed inter-governmental agreements with most of the developed market economies for the promotion of trade and technology transfer. Some of the earliest inter-governmental trade agreements and trade arrangements were signed in the mid 1950s with subsequent agreements on co-operation in science and technology in the late 1960s. Since the early 1970s, however, many of the areas covered by the previous trade and scientific and technological exchange agreements have been absorbed into inter-governmental agreements for economic, scientific, industrial and technological co-operation. These inter-governmental agreements can be considered as 'umbrellas' for trade and technology transfer to be conducted, acknowledging formal government support for the companies, foreign trade enterprises, and other organisations engaged in the actual trade and technology transfer activities. The agreements usually have separate sections which relate to 'scientific co-operation', 'technological co-operation' and also to 'industrial co-operation'. For the purposes of this book, however, attention will be paid to technological and industrial co-operation only, since these are chiefly concerned with the industrial application of scientific and technological information.

To the present author's knowledge, there has been no detailed analysis of the topics covered by each of the inter-governmental agreements between the socialist countries of Eastern Europe and the developed market economies, although a study of the inter-governmental agreements between UK and Bulgaria, Romania, GDR and USSR (44), reveals certain listed 'promising areas of co-operation' which were common to all, or most, of the socialist countries. These included the following:-chemical industry, construction industry, food and food processing, machine tools, mechanical handling and materials handling, metallurgical industries and office machinery. Other 'promising areas' were specific to only one country, usually Romania, and included the aircraft industry, animal husbandry, the automotive industry, environmental protection, and nuclear energy. In addition some of the definitions of 'promising areas' were very general (e.g. 'light industry' for Bulgaria and Romania), whereas others were more specific in relation to the actual sector of light industry (e.g. 'shoemaking' and 'textiles' in the case of the GDR).

BRITISH TRADE PERFORMANCE

The published international trade statistics also reveal certain back-ground information on British exports to the socialist countries of Eastern Europe, compared to exports to that region by the developed market economies as a whole. As discussed above, the developed market economies have enjoyed a trade surplus with the socialist countries since 1971 (see Table 2.2); this surplus accounting for some 8 per cent of total trade turnover from 1972 to 1978, with both the USSR and the remaining Eastern European countries (45). British trade with the USSR, however, has shown a substantial deficit (see Table 2.10), accounting for more than 30 per cent of total trade turnover from 1972 to 1978. Consequently, even though the United Kingdom enjoyed a trading surplus of some 5 per cent of total trade turnover with the remaining East European countries during that time interval, the deficit with the USSR led to an overall level of deficit for British trade with the region of some 12 per cent of total trade turnover (46).

In addition, the share of the developed market economies' exports to the region accounted for by the UK has been in decline. In 1963, for example, UK exports accounted for almost 11 per cent of the developed market economies' exports to the socialist countries, but this proport-ion had fallen to 8.6 per cent by 1970 and 5.4 per cent by 1978 (47).

A decline in the British market share of Western exports of engineer-ing products to that region has also been apparent (see Table 2.11), during the time that the socialist countries have presented themselves as growth markets whilst their engineering factories have been expanded and refurbished (see Table 2.3). In 1963, the UK held over 15 per cent of the share of the developed market economies' exports of engineering goods to that region, but this share had declined to less than 11 per cent by 1970, and less than 6 per cent by 1978 (48). Further study of relevant international trade statistics reveals that the British eng-ineering industry had not only lost its market share in that region to its traditional Western European competitors (Federal Republic of Germany, France and Italy), but also to the USA, a comparative newcomer in that market area. Of particular importance is the poor British per-formance in the Soviet market for engineering goods (see Tables 2.9 and 2.12) where the UK share of the developed market economies' exports fell from 8.5 per cent in 1971 to 4.5 per cent in 1978 (49). Exports from Italy, France and the Federal Republic of Germany appear to have account-ed for a large proportion of lost British sales to the USSR in the late 1960s and early 1970s, with these countries being joined by Japan and the USA in the late 1970s (see Table 2.12).

When the exports of engineering goods to the socialist countries by the UK together with its major Western competitors (50) are investigated in more detail, it can be seen that they show a different structure, when compared with their exports of engineering goods to the world as a whole. In 1975, for example, 'transport equipment' (SITC Group 73) accounted for only some 17 per cent of engineering exports to the

Table 2.10

UK Trade with the Socialist Countries of Eastern Europe (1958-79)
(All figures in millions of $US FOB)

Year	Exports to SCEEs	Imports from SCEEs	Balance	Exports to USSR	Imports from USSR	Balance	Exports to E. Europe	Imports from E. Europe	Balance
1958	130	275	-145	66	145	-79	64	130	-66
1959	170	330	-160	77	165	-88	93	165	-72
1960	215	390	-175	105	190	-85	110	200	-90
1961	295	440	-145	120	225	-105	175	215	-40
1962	310	430	-120	120	215	-95	190	215	-25
1963	345	465	-120	155	215	-60	190	250	-60
1964	280	530	-250	105	240	-135	175	290	-115
1965	315	590	-275	125	290	-165	190	300	-110
1966	410	670	-260	140	330	-190	270	340	-70
1967	479	698	-219	177	340	-163	302	358	-56
1968	548	727	-179	249	379	-130	299	348	-49
1969	553	797	-244	233	473	-240	320	324	-4
1970	596	608	-12	234	266	-32	362	342	20
1971	591	625	-34	206	277	-71	385	348	43
1972	654	687	-33	216	271	-55	438	416	22
1973	762	893	-131	230	363	-133	532	530	2
1974	994	1,128	-134	256	476	-220	738	652	86
1975	1,294	1,492	-198	466	867	-401	828	625	203
1976	1,178	1,871	-693	431	1,196	-765	747	675	72
1977	1,457	2,190	-733	606	1,374	-768	851	816	35
1978	1,872	2,277	-398	812	1,328	-508	1,060	950	110
1979	2,061	2,966	-905	891	1,760	-869	1,170	1,206	-36

Sources: 1968 UN Yearbook of International Trade Statistics, Table B.
1970-71 UN Yearbook of International Trade Statistics.)
1975 UN Yearbook of International Trade Statistics.)UK country tables.
1979 UN Yearbook of International Trade Statistics.)

Table 2.11

Western Exports of Engineering Products to the SCEEs (1963-79)
($US Millions FOB)

Year	UK	France	Italy	FRG	Japan	US	Total for 6 Western nations
1963	154	68	95	189	86	5	
1964	120	82	87	262	132	8	
1965	114	82	93	270	82	11	652
1966	203	124	128	329	120	25	
1967	204	207	179	438	98	31	
1968	282	264	209	445	71	33	
1969	244	288	310	510	95	66	
1970	248	311	323	483	126	69	1560
1971	240	379	290	546	170	89	
1972	274	417	303	997	299	103	
1973	303	466	402	1599	333	270	
1974	327	648	528	2079	422	424	
1975	573	1259	844	2784	835	767	7062
1976	475	1232	702	2815	918	763	
1977	532	1111	978	3147	1181	554	
1978	706	1239	991	3524	1548	585	
1979	707	1659	1002	3160	1155	544	8227

Source: Economic Commission for Europe; *Bulletin of Statistics on World Trade in Engineering Products*; published annually.

Table 2.12

Western Exports of Engineering Products to the USSR (1963-79)
($US Millions FOB)

Year	UK	France	Italy	FRG	Japan	US	Total of 6 Western nations
1963	61	27	48	80	76	1	
1964	43	32	38	129	116	5	
1965	50	27	32	80	60	5	254
1966	75	32	31	82	88	4	
1967	76	78	50	83	49	11	
1968	129	152	72	116	46	15	
1969	113	145	174	200	70	42	
1970	101	142	174	164	103	45	729
1971	84	139	140	169	117	63	
1972	96	119	128	367	196	62	
1973	104	169	154	575	168	204	
1974	69	290	172	759	230	225	
1975	210	574	390	1350	558	547	3629
1976	205	411	358	1397	682	605	
1977	191	577	559	1504	769	374	
1978	299	635	501	1537	1168	317	
1979	287	821	497	1399	762	363	4129

Source: Economic Commission for Europe; *Bulletin of Statistics on World Trade in Engineering Products*; published annually.

socialist countries, whereas this group accounted for some 38 per cent of the selected Western nations' total exports of engineering goods (51). The socialist countries of Eastern Europe can consequently be considered primarily as 'capital goods' markets, particularly of 'machinery non-electric' (SITC Group 71). In 1975, that product group accounted for 69 per cent of the imports of engineering goods by the socialist countries from the selected Western nations (73 per cent in the case of the USSR, and 66 per cent for the remaining Eastern European countries) compared with 43 per cent for the world as a whole (52). In that same year, the socialist countries of Eastern Europe received almost 7 per cent of the total exports of 'machinery non-electric' from the selected Western countries, 3.7 per cent being delivered to the USSR, and 3.2 per cent being delivered to the remaining socialist countries (53). For this product group, however, the UK market share in the socialist countries of Eastern Europe, compared to its Western competitors, fell from 22 per cent in 1965, to 18 per cent in 1970, and 8 per cent in both 1975 and 1978. For exports to the USSR, the corresponding figures were 28 per cent, 16 per cent, 5 per cent and 7 per cent respectively (54).

There is little reliable evidence on British success in the sale of licences to the socialist countries of Eastern Europe. Evidence on industrial co-operation activity should also be treated with a certain amount of caution, in view of the lack of any comprehensive official statistics and the difficulties encountered by independent researchers in attempting to compile this data (55). As stated in a previous section however, data available from the United Nations Economic Commission for Europe (ECE), suggests that British companies were in eighth place in terms of the quantity of industrial co-operation agreements signed with the socialist countries following Western Germany, France, Austria, USA, Japan, Italy and Sweden (56).

NOTES

(1) See for example, Wilczynski (1969), Askansas *et al.* (1978) and Levcik (1978).
(2) i.e. the *Yearbook(s) of International Trade Statistics* which are published annually by the United Nations Statistical Office, and also the *UN Monthly Bulletin(s) of Statistics*.
(3) i.e. those Eastern European countries which are members of the Council for Mutual Economic Assistance (CMEA or COMECON), namely Bulgaria, Czechoslovakia, German Democratic Republic (GDR), Hungary, Poland, Romania, USSR. United Nations statistics also include foreign trade for Albania within this group for certain years, but its volume is sufficiently small to enable it to be neglected.
(4) i.e. the developed market economies of Western Europe, including Greece and Yugoslavia, USA, Canada, South Africa, Australia and Japan.
(5) See Marer, P. 'Toward a Solution of the Mirror Statistics Problem in East-West Commerce' in Levcik (1978) for a further discussion of this problem.
(6) Special Table B (World Trade of Market Economies: Index Numbers by Regions) of the January 1980 *UN Monthly Bulletin of Statistics* provides the following information on the Unit Value (Price) Index (in US dollars) for World Exports from the Market Economies:-

1960	1965	1968	1969	1970	1971	1972	1973	1974	1975	1976	1977	1978
40	42	43	44	46	49	53	66	93	100	102	111	122

Some of these increases in the unit value (price) index would be due to general technological and quality improvements, in addition to the major factor of inflation.

(7) See Hill (1978), pp.6-22,85-135.

(8) The source referred to in note (6) above provides the following information on the Unit Value (Price Index (in US dollars) for the World Exports from the Developed Market Economies.

1960	1965	1968	1969	1970	1971	1972	1973	1974	1975	1976	1977	1978
45	47	48	49	52	55	59	72	89	100	100	109	123

(9) See Askansas, Fink and Levcik (1978) for a discussion of the CMEA debt using Eastern European data sources; and also US Congress Office of Technology Assessment (1981), pp.37-50 for a discussion of the size of the Eastern debt, and its possible effects on East-West trade.

(10) See, for example, Wasowski (1973). Young (undated) has also attempted to quantify trade in technology even more accurately using classes of 'high technology goods' in US exports. Zaleski and Wienert (1980) also give a detailed contemporary survey of available published data on East-West trade and technology transfer.

(11) See Wilczynski (1976a), p.8. The material in that paper was subsequently published in Wilczynski (1977).

(12) See Wilczynski (1977), p.126.

(13) *Ibid.*, p.130.

(14) *Ibid.*, p.131.

(15) *Ibid.*, Wilczynski quoted R. Osterland, *Die Wirtschaft*, East Berlin, 1972, No.37, p.27 as his source.

(16) See Wilczynski (1977), p.131.

(17) *Ibid.* Hanson (1976) also provides various estimates of CMEA licence transactions with the developed West.

(18) The Chase World Information Series on East-West Business Co-operation and Joint Ventures in its volume on *Hungary* (NY, 1976), p.61 quoted the following data on Hungary's receipts and payments for licence transactions with the West, citing *Kulkereskedelmi Statistikai Evkonyv, 1968-1974* (all figures in millions of $US).

Year	Receipts from West	Payments to West	Deficit
1968	0.17	0.94	0.78
1969	0.29	0.54	0.26
1970	0.44	1.49	1.05
1971	0.73	6.72	5.99
1972	0.94	5.25	4.31
1973	2.99	3.61	0.63

(19) See Kiser (1977) and Kiser (1980).

(20) See ECE (1979), pp.23,24.

(21) See Bykov (1977).

(22) See Szita (1977), p.163.

(23) See Levcik and Stankovsky (1979), p.173.

(24) See ECE (1976b).

(25) See McMillan (1977a), pp.1184-1187.

(26) See Wilczynski (1976b), p.80.

(27) See Szita (1977), p.163.

(28) See McMillan (1977a), p.1207.

(29) See ECE (1976c) and Kemenes (1974).

(30) See Szita (1977), p.174.

(31) See McMillan (1977a), p.1210, and Tabaczynski (1974).

(32) See Olszynski (1973).

(33) See Levcik and Stankovsky (1979), p.176.

(34) See Scott (1978) and Knirtsch (1973).

(35) See McMillan (1977a) p.1183 for a further discussion of this point.
(36) See ECE (1973).
(37) See ECE (1976b and 1978a).
(38) See McMillan (1977a), p.1186.
(39) See ECE (1976b).
(40) See ECE (1976b and 1978a).
(41) See McMillan (1977a), p.1188.
(42) See McMillan (1977a), p.1186.
(43) See Theriot (1976).
(44) See the following inter-governmental agreements for co-operation in the fields of science, technology and industry:- Anglo-Bulgarian Agreement, 14 May 1974; Anglo-Romanian Agreement, 18 September 1975; GB-GDR Agreement, 18 December 1973; GB-USSR Agreement, 6 May 1974.
(45) From 1972-1978 DMEs Exports to SCEEs $167,674m

$$\text{SCEEs Exports to DMEs} \qquad \$141,955m$$
$$\text{Therefore total turnover} = \$309,629m$$
$$\text{SCEE Deficit} = \$\ 25,719m$$
$$\text{Therefore } \frac{\text{Deficit}}{\text{Total Turnover}} = \frac{25,719}{309,629} = 8.3\%$$

$$\text{Similar ratios for USSR} \qquad \frac{11,716}{145,368} = 8.1\%$$

and for the remaining socialist countries $\dfrac{14,016}{164,008} = 8.5\%$

(Compiled from data shown in Table 2.2).
(46) Using a similar method to that outlined in note (45) above, for the UK from 1972 to 1978.

$$\frac{\text{Deficit}}{\text{Total Turnover}} = \frac{2,320}{18,742} = 12\% \text{ (UK deficit with SCEEs)};$$

$$\frac{2,850}{8,884} = 32\% \text{ (UK deficit with USSR)};$$

$$\frac{530}{9,858} = 5\% \text{ (UK credit with remaining socialist countries).}$$

(Compiled from data shown in Table 2.10).
(47) From data shown in Tables 2.2 and 2.10, UK market shares of DME exports to the SCEEs were as follows:

$$1963 \quad \frac{345}{3,170} = 10.9\%$$

$$1970 \quad \frac{596}{6,939} = 8.6\%$$

$$1978 \quad \frac{1,872}{34,453} = 5.4\%$$

(48) From data shown in Tables 2.3 and 2.11, UK market shares for DME exports of engineering goods to the SCEEs were as follows:

$$1963 \quad \frac{154}{1,000} = 15.4\%$$

$$1970 \quad \frac{248}{2,381} = 10.4\%$$

$$1978 \quad \frac{706}{12,670} = 5.6\%$$

(49) From data shown in Tables 2.9 and 2.12, UK market shares for DME exports of engineering goods to the USSR were as follows:

$$1971 \quad \frac{84}{983} = 8.5\%$$

$$1978 \quad \frac{299}{6,630} = 4.5\%$$

(50) i.e. France, Italy, Federal Republic of Germany (FRG), USA and Japan.

(51) The export figures shown below have been compiled from the *Bulletin of Statistics on World Trade in Engineering Products - 1975* (published by the Economic Commission for Europe, N.Y.) for UK, France, Italy, FRG, USA and Japan. They relate to the total world exports for these selected Western countries, and also their exports to the SCEEs.

	Total World Exports	Exports to the SCEEs
Exports of Transport Equipment (SITC Group 73)	$ 62,427.7m	$ 1,197.5m
Exports of Engineering Goods (SITC Group 7)	$163,942.4m	$ 7,063.1m
Transport Equipment as % of Engineering Exports	$\frac{62,427.7}{163,942.4} = 38\%$	$\frac{1,197.5}{7,063.1} = 17\%$

(52) The export figures shown below have been compiled from the same source as in note (51) above, for 1975 (i.e. for UK, France, Italy, FRG, USA and Japan).

	Exports to the SCEEs	Exports to the USSR	Exports to the remaining 6 SCEEs*	Total World Exports
Exports of 'Machinery non-electric' (SITC Group 71)	$ 4,893.0m	$2,635.2m	$2,257.8m	$70,732.5m
Exports of Engineering Goods (SITC Group 7)	$ 7,063.1m	$3,628.8m	$3,434.3m	$163,942.4m
'Machinery non-electric' as % of Engineering Exports	$\frac{4,893.0}{7,063.1}$ $= 69\%$	$\frac{2,635.0}{3,636.8}$ $= 73\%$	$\frac{2,257.8}{3,434.3}$ $= 66\%$	$\frac{70,732.5}{163,942}$ $= 43\%$

* i.e. Exports to the SCEEs - Exports to the USSR.

(53) i.e. for 'machinery non-electric' (SITC Group 71), in 1975.

	Selected Western Countries*	%
Exports to USSR	$ 2,635.2m	3.7
Exports to 'Remaining SCEEs'	$ 2,257.8m	3.2
Total World Exports	$70,732.5m	100.0

* i.e. UK, France, Italy, FRG, Japan, USA.

(54) For 'machinery non-electric' (SITC Group 71).

	UK exports to SCEEs	'Selected Western Countries'* exports to SCEEs	UK exports as % of Western exports to SCEEs	UK Exports to USSR	'Selected Western Countries' exports to USSR	UK exports as % of Western exports to USSR
1965	$ 91.4m	$ 420.8m	22	$ 43.5m	$ 154m	28
1970	$198.6m	$1,115.6m	18	$ 83.3m	$ 529.4m	16
1975	$406.8m	$4,893.1m	8	$134.9m	$2,635.3m	5
1978	$535.4m	$6,508.4m	8	$232.3m	$3,435.5m	7

* i.e. UK, France, Italy, FRG, Japan, USA.

Source: Compiled from data shown in *Bulletin of Statistics on World Trade in Engineering Products* for 1965,1970,1975 and 1978.

(55) Almost all of the compilers of statistics relevant to industrial co-operation express reservations about the accuracy of their data. With regard to the sale of licences, it is not usual for statistics to be published officially in great detail for sales of this industrial property to the socialist countries. Furthermore, it is frequently the case that available data does not include sale of licences as part of a larger project (e.g. purchase of equipment).

(56) See ECE (1976b).

3 East-West trade and technology transfer — the case of Soviet imports of British machine tools

INTRODUCTION

Since the initial establishment of a machine tool industry in the early
1930s, the USSR has developed a strong base for machine tool design and
production, as evidenced by the achievements of the Soviet engineering
industry over the past forty years. By the mid 1960s, the Experimental
Scientific Research Institute for Metalcutting Machine Tools in Moscow,
together with its experimental factory ('Stankokonstruktsiya') were
capable of developing and manufacturing many advanced types of machine
tool, whilst the large number of specialised design offices ('SKB's')
and drawing offices in individual factories had also assimilated quite
high levels of design expertise. (1) Furthermore, although the majority
of Soviet machine tool factories specialised in the large batch manufacture of general purpose machine tools (2), many producers also had
the capability to manufacture the more technologically sophisticated
types of machine (3); and certain user industries and factories had the
requisite expertise to design and build some of the advanced machines
which they required for their own use. (4)

   In spite of these developments in Soviet machine tool design and production capacity, however, there is evidence to suggest that the industry
still experienced the following problems during most of the 1960s and
1970s:

(a) a lag behind its Western counterparts in terms of its capability to
    produce machine tools of similar accuracy, dynamic rigidity and
    reliability (5);
(b) insufficient design and production capacity to meet the high demands
    of certain large end-user industries (e.g. the automotive industry)
    for special purpose machine tools, transfer machines and link lines
    (6). Furthermore, problems in the technical development of these
    types of equipment (7) may have been heightened by administrative
    barriers between industrial ministries tending to limit the develop-
    ment and supply of requisite specialised items such as hydraulic,
    pneumatic and control devices ('proprietary items');
(c) a comparatively late entry into the development and batch production
    of numerically controlled machine tools, caused by administrative
    barriers between control equipment suppliers and the machine tool
    industry, and shortcomings in the availability of appropriate
    computer hardware and software (8).

It is important to bear in mind, however, that if analogous critical surveys had been carried out on machine tool industries in each of the advanced Western industrial nations, it is probable that technological lags and shortfalls in production capacity would also have been found for particular sectors of many individual national industries, as a result of specialisation (9) - a factor which influences the comparatively high level of international trade in machine tools amongst the advanced industrial nations. Furthermore, it would probably have been economically inefficient for any national industry to attempt to achieve complete leadership in every field of machine tool technology, and the USSR is no exception to this general rule in spite of the attempt of Soviet leaders to follow patterns of autarky and rapid technological advance for political reasons. In such circumstances it was consequently realistic for Soviet industry to have attempted to absorb certain aspects of machine tool technological know-how from advanced Western industrial nations in those important fields where it was technologically lagging, and where it would have been impractical to develop domestically the technology or requisite production capacity, in view of other tasks to be undertaken.

Machine tool technology can be absorbed through a number of avenues (10), including:-

(a) study of relevant technical and trade literature;
(b) participation in technical seminars;
(c) discussions on machine tool technology with technically advanced makers and users of machine tools;
(d) purchase of licences;
(e) co-operation in research and development;
(f) purchasing of machine tools, and integrated sets of special purpose machinery.

This author considers that avenues (d), (e) and (f) above have provided the best channels for the absorption of machine tool technology from the West, by the USSR. As illustrated in a previously published study (11), there is evidence to suggest that imports of machine tools from Western countries accounted for more than 10 per cent of Soviet consumption of machine tools during the 1971-75 period, although there is little comprehensive information on purchasing of licences or co-operation in research and development, at that time. The greatest incentive to the USSR to import on this scale is likely to have been the securing of production capacity, followed by the subsequent experience gained in the day to day operation of the actual machines supplied; especially when the expansion of a particular user industry (the automobile industry) was occurring at too fast a rate for domestic suppliers to develop and produce the necessary equipment.

To the Western exporter, the USSR was a large market at that time, with approximately 10 per cent of machine tool exports from some advanced Western countries being delivered to the USSR during 1971-75 (12); and most Western exporters would consequently have been likely to meet most Soviet requests for relevant technical information, in view of perceived sales possibilities in that market following the provision of such data.

The studies to be described in this chapter are consequently concerned with an evaluation of the manner in which the USSR has attempted to absorb Western machine tool technology into its own engineering industry through its foreign trade process, during recent years. Particular attention has been paid to the technology transferred, and the time scales over which the technology was absorbed into Soviet industry.

In order to obtain information on this topic, it was decided to interview British business executives with experience in the export of machine tools to the USSR, since these executives were considered to be one of the main available sources of information on the operation of this Soviet technology absorption mechanism. Furthermore, British companies had been important suppliers of machine tools to the USSR, accounting for 30 per cent of Soviet imports of Western machine tools in 1965 and 1966; although this share fell to 15 per cent in 1971, and 3 per cent in 1975 (13).

Several companies were approached by means of an introductory letter, chiefly those engaged in the design and production of machine tools for use in the Soviet motor industry, since available Soviet and British foreign trade statistics suggested that this industry accounted for a large proportion of Soviet machine tool purchases from Britain during the late 1960s and the 1970s (14). In those cases where a positive response was received, a questionnaire was forwarded to the company to form a standardised structure for discussion at a subsequent interview (see Appendix A). The topics discussed during the interviews conducted during late 1978 were categorised as follows:

technical background,
proposal and contract,
acceptance and installation,
utilisation, and
diffusion.

The eight companies which provided the information can be categorised as follows:

two designers and producers of special purpose transfer machines and link lines for automotive components;
a designer and builder of special purpose gearcutting machinery;
a designer and producer of numerically controlled machine tools;
two designers and producers of automatic turning machines;
a designer and producer of precision grinding machines;
a designer and producer of automotive components and of associated production equipment.

The following section of this chapter is an account of these companies' experience of the Soviet absorption of machine tool technology.

CASE STUDIES OF SOVIET IMPORTS OF BRITISH MACHINE TOOLS

Case study no. 1 - a producer of special-purpose machining systems

Technical background. The company which provided information for this case study is a designer and producer of special purpose machine tools, transfer machines and link lines, chiefly used for the repetitive manufacture of automotive engine and transmission components. During the late 1960s and the early 1970s the company signed a series of contracts with the USSR for the supply of specialised machine tools and transfer

lines for the production of a range of automotive components. The total
value of these sets of equipment, which were to be installed in three
separate Soviet factories, was some £5 million. Since their successful
completion, the company has also received further orders for special
purpose machines for carrying out individual operations on other types
of automotive component.

For the purpose of this research, however, it was decided to concentrate on two transfer line projects only, as both of these projects were
of a turnkey (15) type of operation and both projects also followed the
company's general experience with regard to the proposal and contract
stages of Soviet projects. In the one case the plant was installed at
an established factory whilst in the other case the plant was installed
at a greenfield site, thereby providing some comparative information at
the acceptance and installation stages.

The main technical features of that equipment provided by the company
can consequently be summarised as the capability to automatically carry
out the sequence of operations necessary to completely process large
batches of identical automotive components from casting to finished
component, and also to automatically sequentially transfer the components
through the relevant workstations. None of the technical features of
the equipment were patented by the company.

The company was of the opinion that the USSR, at that time, was not
able to build machines of the same reliability, or of the capability to
produce components repetitively to the required accuracy over a complete
working shift. The company considered, however, that technology transfer factors were probably secondary to those of domestic capacity shortfalls, when the USSR decided to import such equipment at that time.

The Soviet customer was obliged to carry out process development work
before the imported machinery could be used satisfactorily. At the
greenfield site, the castings as originally presented for machining were
completely unsatisfactory, although this problem was resolved when a new
imported casting machine became operational.

Proposal and contract. The time intervals which elapsed between receipt
of initial enquiry to placing of order were found to be quite lengthy
for the selected Soviet contracts (between one and two years) compared
with the company's experiences in Western markets (six months approximately). This interval was considered to have been escalated through
difficulties in contacting the end-user over technical matters, as a
consequence of the necessity of channelling all communications through
the foreign trade organisation.

The initial enquiry as received by the company was usually quite
sketchy, and three months usually elapsed before the company had sufficient information to submit an adequate proposal. A series of subsequent
modified proposals were usually submitted over some one to two years
before a contract was signed, and it was usually the case that the final
proposal on which the contract was based differed substantially from the
original submission. In some cases this caused costs to escalate as the
final proposal was more technically sophisticated than the original
proposal, although in other cases costs were sometimes reduced when the
Soviet customer decided to take only one part of the proposed total
machining section.

It is important to note that the type of proposal submitted by the company usually requires close rapport with the customer, since the company usually includes process planning proposals as part of its overall technical proposal. When carrying out such process planning, it is usually found necessary to have quite frequent meetings with the customer's production engineers in order that common experience in the manufacture of the particular type of component can be pooled. In the case of the USSR, however, it was found that there were lengthy time intervals between these meetings thus causing the whole cycle of submission and acceptance of proposals to be lengthened. The process was also influenced by the large amount of detailed technical information usually required by the Soviet customer, and the tendency for Soviet buyers to wish to formalise contractually, more technical aspects than are found to be usual in Western markets.

Acceptance and installation. The company has found this stage to be especially lengthy on Soviet contracts as a result of a number of factors. In the first place, it was generally found that the initial machine design engineering stage was delayed quite substantially as a result of lack of contact between the company and the end-user. In Western markets it is usual to have frequent meetings (sometimes weekly) between the machine designers, or salesmen, and the customer, in order that queries can be quickly answered, and common experience pooled in the interests of good design. This was found to be impossible when dealing with the Soviet customer, however, as a result of difficulty in communications between the three parties (i.e. the supplier, the foreign trade organisation, and the user). These time lags in communication consequently caused increased times for the three to four month stage of design engineering.

Secondly, it was found that difficulties were encountered during the inspection stages at the company's factory, partly as a consequence of the apparently sketchy technical briefing of the Soviet inspectors before their visit to the UK and also partly as a consequence of inspectors sometimes being reluctant to make decisions. These difficulties were additional to the normal Soviet business practice of refusing to deviate from technical contractual conditions even though these deviations may have been of negligible importance (e.g. insistence that each work station performed exactly to the stage drawing documented at the process planning proposal stage, in addition to the finished component being produced to specification).

As a consequence of these delays in engineering and inspection it was found that a normal planned design and build time of twelve months could be increased to eighteen months for an established site, and up to two years for a greenfield site.

The final factor influencing acceptance delay, and perhaps the most important factor, was found to be the Soviet approach to installation work, particularly in new factories. The company considered that installation work for similar machines would have taken some six months in Western factories, with engineers calling in for minor adjustments and modifications for an additional six months. In established Soviet factories it was estimated that installation and commissioning would take some twelve months, whilst at a greenfield site installation and commissioning could take up to three years. Several reasons were advanced for this delay, namely: the complexity of the equipment; frequent shortages in the on-site availability of suitably trained manual and technical labour; the time scales required to achieve co-ordination between relevant Soviet organisations involved in the project; and the quality of raw materials as presented to the machine.

Total time from enquiry to installation. The time interval between enquiry and final acceptance was consequently found to take some five to six years. This was approximately two to three times the usual time interval encountered with Western European customers.

Utilisation. The company had no information on the utilisation of the equipment following installation, but considered the level of manning to be generally higher in the USSR than in the West.

Diffusion. The company did not know of any particular cases where the basic principles or technical refinements of its equipment had been incorporated into Soviet-built equipment, and considered identical replication to be almost impossible because of lack of volume supply of such proprietary items as hydraulic pumps and valves, and precision tooling. However, since the Soviet customer had imported large volumes of these items as spares, these may have been subsequently used along with Soviet designed machine elements in other equipment.

The company wondered whether, even if relevant machine design features of Western equipment were to be incorporated, the tradition of machine tool assembly in the USSR was sufficiently developed to manufacture machinery having the requisite high production performance demanded by contemporary Soviet volume production industries.

Case study no. 2 - a producer of numerically controlled machine tools

Technical background. The company interviewed for this case study is a designer and builder of general purpose and numerically controlled machine tools. The information contained in this case study refers to one Soviet export contract only, namely the sale, during the mid-1970s, of a vertical-spindle turret-type machining centre equipped with a control system capable of both point-to-point and contouring. The machine was specially tooled for milling, drilling, tapping and reaming operations on a particular component to be produced in the end-user's factory.

The company was of the opinion that, although the Soviet engineering industry at that time may have been capable of producing a machine having a similar overall specification, it was doubtful whether the machine would have the same compactness, accuracy and reliability as the company's product. This was considered to be mainly due to the many years of design and production expertise of the company, and also due to the presence of highly developed Western industries capable of the specialised manufacture of necessary components and assemblies for use in the machine tool industry (e.g. recirculating ball screws and rolling bearing units for slideways). The USSR did not appear to have such a highly developed proprietary item industry at that time, and Soviet numerical control technology was also apparently experiencing problems at the control interface stage. It was estimated, therefore, that the Soviet machine tool industry was some five years behind its Western counterparts in the design and construction of such equipment, and that the equipment was imported to meet an accurate, highly-productive capacity requirement in the end-user's factory.

It was considered that the end-user would have had to carry out little process development work to accommodate the machine – numerically controlled machinery was in quite extensive use in the factory, which was also believed to have had its own computer for a number of years.

Proposal and contract. The company received the relevant enquiry from a Soviet foreign trade organisation during the autumn of a certain year. A proposal was forwarded that same month, and subsequently modified one month later to include a control system produced by a different manufacturer, and additional tooling. This caused the price to increase by some 12 per cent compared with the original, and the final value of the machine, control equipment and tooling was in the region of £80,000. A contract was signed with the customer during the mid-summer of the following year, in line with this proposal. The comparatively long time-lag between proposal and signing of contract was considered to be caused by extensive study of competitive quotes by the foreign trade organisation itself, and also relevant discussions between the foreign trade organisation and the end-user.

Acceptance and Installation. The delivery date for the machine tool was during the winter of the year following the signing of the contract (i.e. a design and build time of some eighteen months). The machine was built in some twelve months, although a Soviet inspector did not arrive until some six weeks before the agreed date of shipment. The inspector was trained in the operation, installation, programming and maintenance of the machine over this six-week interval. These tasks usually took some three weeks for similar contracts, but were extended in the Soviet case in view of the larger amount of technical documentation required by the Soviet customer, since the end-user factory was carrying out its own installation.

The machine was successfully accepted by the agreed delivery date and shipped FOB British port accordingly.

Utilisation. The company had no further information about the utilisation of the equipment, since it was not responsible for its installation and commissioning. It was their opinion, however, that it should be utilised as effectively as in a Western factory in view of the back-up facilities within the Soviet factory, and the high degree of technical expertise of the visiting Soviet inspector who was to carry out subsequent commissioning and operation.

Diffusion. The company did not possess any information concerning the incorporation of their equipment's design features into Soviet-built machinery, but doubted Soviet ability to fully replicate the machine tool itself in view of the lack of a fully developed proprietary-item industry at that time. It was likely that the control system could be replicated, but the company considered that Western control industries had now moved into other advanced areas.

Case study no. 3 – a producer of turning machines

Technical background. The company which provided information for this case study is a designer and producer of multi-spindle automatic turning machines. During the late 1960s and early 1970s it delivered a total of almost eighty of these machines, and associated specialised tooling, to a Soviet foreign trade organisation for installation in a new Soviet

passenger car factory; the products and processes of this factory were
designed with the technical assistance of a Western European car manufacturer. Each of the purchased machines was specially tooled for the
repetitive manufacture of a particular item; namely specific transmission, engine, gearbox, or body components. None of the technical features
of the equipment was covered by patents for which the USSR became a
licensee.

The company was of the opinion that the Soviet machine tool industry
was capable of producing quite adequate multi-spindle automatic turning
machines at that time, although the normal commercial tolerances
specified in Soviet state standards (16) for these products tended to be
slack. The tightened tolerances for higher precision multi-spindle
machines, on the other hand, were considered to be too tight for normal
commercial application, and some requirements appeared to be drawn up
by theoretical engineers with little appreciation of commercial practice
(17). The main differences in Soviet and Western practice were considered to be in the area of tooling where Western international competitive
pressures for client cost-reduction caused Western machine tool producers to offer continually improved machining packages to their clients,
providing improved component precision together with reduced production
cycle times. The company considered these competitive pressures to be
absent in the Soviet Union, which consequently caused a Soviet technological lag in this important area of machine tool application. The
company found it to be impossible to quantify this technological lag,
however.

The company considered the major Soviet reason for importing its products to be the Soviet decision to rapidly expand its passenger car production during the late 1960s. The general backwardness in Soviet automotive design and production techniques at that time made it necessary
to purchase design and production know-how from a Western company, if
the USSR was to attempt to meet its production targets for contemporary
models of passenger car. This backwardness, in its turn, caused the
USSR to require imported machines capable of carrying out the production
operations at the quality, reliability and rate of production specified
by the Western automobile producer, based upon experience in his own
factory.

The main area of technical development considered to be required by
the USSR to accommodate the company's machinery was in the area of
improved availability of free-machining steels. The specification of
Soviet bar stock was found to be different from that usually encountered
in Western countries from the viewpoint that although Soviet steels were
found to be completely adequate in terms of their mechanical properties
(e.g. tensile strength), they were found to be limited in terms of their
machinability during component production. In view of the fact that the
USSR was purchasing large quantities of Western machinery to achieve
Western productivity levels in selected Soviet motor factories, it would
have been necessary for the Soviet steel industry to develop steels
having machinability qualities capable of enabling the imported machinery
to achieve the requisite component production rates.

Proposal and contract. The company received an enquiry from the Soviet
foreign trade organisation's European branch office at the end of a
certain year in the late 1960s. It subsequently submitted its proposals
during the early part of the following year, approximately one month

after receipt of the enquiry. During the autumn of that same year discussions were commenced with the Soviet foreign trade organisation and subsequently with production engineers from the Western car company which was providing technical assistance. As a result of these discussions, the three sides pooled their experience towards a final technical proposal. During the early part of the following year, the commercial negotiations were commenced in earnest, and two contracts were subsequently signed in the early summer of that same year, of total value of some £1.2m (1969 prices). The final technical proposal was not considered by the company to differ substantially from its original proposal although the number of machines was slightly less for similar quality and output, as a result of improvements in tooling and other aspects of machining technology.

The time that elapsed between first proposal and signing of contract was consequently some eighteen months, and the company advanced several reasons for this. The first was the way in which enquiries were initially launched by the foreign trade organisation - there was a 'floodgate effect' caused by a large number of enquiries being released simultaneously. The foreign trade organisation would consequently have been inundated with a large volume of quotations from a number of companies - the detail of a typical proposal required from the company was such that one complete top copy of the quotation fitted into a box of approximately two cubic metres in volume! Similar proposals for the manufacture of the same components were also received from some half-dozen of the company's competitors, although this figure was subsequently reduced to two. The Soviet buyers would consequently require a long time to evaluate such a large volume of detailed proposals. The second feature was the length and detail of the technical discussions which were carried out between three parties: the company, Soviet engineers, and engineers from the Western car firm providing technical assistance. The final feature was the prolonged nature of Soviet negotiation procedures to obtain the best possible commercial and contractual conditions to the Soviet side, particularly at a time when the Western market for specialised machine tools was generally depressed.

Acceptance and installation. The contractual dates for the delivery of the machines extended from some six months after the signing of the contract to some two years after signing, in view of the large size of the order. It was difficult for the company to be specific when answering the question concerning the meeting of delivery dates in view of the long time interval over which deliveries were scheduled - the majority of machines were delivered on time whilst others were delivered some six to twelve months late. Various difficulties, in addition to the overall size of the project, were mentioned by the company in relation to the meeting of delivery dates, namely:

(a) certain 'grey areas' in contractual conditions which the Soviet side usually attempted to interpret to their own advantage: one example given was the quantity and detail of technical documentation that was required;
(b) frequently inconvenient docking dates for Soviet ships into a British port, and delivery was to be FOB Soviet ship;
(c) delays in inspection and acceptance of machines due to Soviet inspectors tending to initially favour those testing procedures laid down in Soviet state standards; changes in Soviet inspectors over the long delivery time interval; a tendency for Soviet inspectors to refer

frequently to the Trade Delegation of the USSR in London for decisions; and delays in the delivery of testing materials in adequate quantities and quality, partly due, in the company's opinion, to poor co-ordination between the Soviet steel and automobile industries. This problem was further compounded by lower machinability characteristics of the Soviet bar stock, compared with the bar stock used in the Western car company providing technical assistance to the USSR, since bar stock to the Western specification had been used as the basis for pre-contract proving of tooling. Consequently, modifications to the tooling were frequently required to enable the requisite quality and output rates to be achieved.

With regard to the installation of machines on site, it was found that many machines were installed and accepted readily, whilst others were delayed by some six months beyond the date originally expected. The complete installation took some eighteen months in view of the large number of machines, and in some aspects, conditions on site were reasonably well organised – on arrival it was found that conveyors, swarf removal and centralised coolant systems were installed, adequate cranage was available, and there was a resident team of inspectors and interpreters. On the other hand, the factory was far from being built, and there were no back-up facilities available for such things as machining and hand tools. Furthermore, it was almost impossible to obtain rapid decisions for the solution of problems occurring during installation and commissioning, possibly due to poor on-site co-ordination. These problems, in the company's view, were not uncommon on greenfield projects in other Eastern European countries, but were considered to be more typical of problems encountered in Third World projects than in projects in advanced industrialised nations.

Total time for initial enquiry to installation. In view of the size and complexity of the project, the company found it difficult to compare the time that elapsed with that which could be expected from other Western European customers. This time was substantially extended, however, as a consequence of the reasons given above.

Utilisation. A representative of the company had recently paid a visit to the factory and considered the machines to be fully utilised at a level of manning which was no higher than in Western automotive companies carrying out similar operations. Furthermore, it was the representative's view that the machines were adequately maintained.

Diffusion. It was the company's view that the USSR made no deliberate approach to copy its products, although it would be expected that they would assimilate ideas of interest. As previously mentioned, an important aspect of the company's customer package is in the area of tooling, and it is possible that the Soviet customer may have used similar tooling layouts for the production of analogous components. It was wondered, however, whether pressures for high productivity were sufficiently established in the USSR to encourage Soviet factory engineers to transfer the company's technology in this manner.

Case study no. 4 – a producer of gearcutting machines

Technical background. The company which provided information for this case study is a designer and manufacturer of gearcutting machine tools (shapers, hobbers and planers) and gearcutting tool grinding machines.

The company's product range includes a gearcutting machine which is virtually unique in type, namely a twin-spindle gear shaper which facilitates the simultaneous production of a series of gear wheels having a common axis of rotation. The company has sold a large quantity of its standard machinery to the USSR (e.g. gear hobbing machines and single-spindle shaping machines), and has also sold twin-spindle gear shaping machines, specially tooled and equipped with component transfer equipment. This case study will discuss the latter type of machine, since this was the most recent sale that the company had made at the time of writing this case study: the machinery was specially tooled and equipped for cutting a series of gear-teeth on a truck layshaft at the largest and newest Soviet truck factory. None of the technical features of the machine was covered by any patent for which the USSR became a licensee.

The company was of the opinion that the USSR was not producing similar machinery at that time, but found it impossible to give an estimate of Soviet technical lag, since the company was also of the opinion that it was a Soviet choice of research and development resource allocation. Twin-spindle gearcutters are only used in specialised cases where it is required to produce multi-geared components in large quantities, and the overall demand for these machines is consequently quite small compared with the more standard types of single spindle machine. Furthermore, all of the operations which can be done on a twin-spindle machine can also be done with a series of single-spindle machines, although clearly in a more expensive manner since more machines are required, and the component has to be re-loaded in fixtures for each gear diameter. Consequently, since the field of application for twin-spindle machines was comparatively small, a reliable source of import supply existed, and domestic substitution by single-spindle machines was possible, although at a far lower level of efficiency; the USSR appears to have decided not to allocate resources to the development of twin-spindle machines (18).

The Soviet customers had to carry out technical development in the fields of electrical and hydraulic equipment to conform to the company's requirements. It is likely, however, that much of this equipment was also imported from Western sources of supply, since Western product specifications were selected as the basis for the standards.

Proposal and contract. The company was initially approached in 1970 by a Soviet import agency which had a large office based in Paris, from where it co-ordinated the purchase of production machinery for a large Soviet truck plant. An initial study was made by the company of the gearcutting processes required, and proposals in response to enquiries were forwarded to some four Western companies which were bidding for the position of project leader for gearbox production. In 1972, the Soviet buyer selected one of these companies (a West German company which had previously supplied machinery for the manufacture of gearboxes for Soviet passenger cars) to act as project leader for the complete manufacture of gearboxes at the Soviet truck factory. The total value of this project was some £50m, and forty seven companies were to be co-ordinated by the West German project leader. A revised quote was subsequently submitted by the company in 1972, and the contract was signed in mid-1973 for equipment worth some £500,000. Once the West German manufacturer had been accepted as project leader by the Soviet buyer, the required technical changes were minimal since the company had previously submitted a quotation to that German firm. The majority of negotiations were of a commercial nature, the length of time caused by the size of the contract and usual Soviet practices of highly contended commercial negotiation.

Acceptance and installation. Delivery was scheduled to take place between September 1974 and February 1975, and these dates were adhered to, mainly as a consequence of the company's previous experience in the Soviet market of the factors which can sometimes prolong acceptance (e.g. delivery of technical documentation).

The installation of the machinery, together with training of Soviet labour was part of a separate contract. Installation and training took some twelve months in total, although this twelve month period was spread over an eighteen month interval. The skill levels of Soviet labour on this project were found to be quite high, although this could be partially explained by the high priority of the project from the Soviet viewpoint, and the fact that the director of the Soviet truck enterprise was also a vice-minister in the Soviet Ministry of the Automobile Industry. The installation was further facilitated by the on-site presence of a permanent representative of the West German project leader, who co-ordinated installation and acceptance work, ensuring the delivery of correctly pre-processed blanks for manufacture.

The only major problem encountered at the on-site acceptance stage was caused by difficulties in machinability of the castings as produced in the Soviet truck factory, which differed in machining properties from the castings delivered (probably from another source) for machine testing in the company's factory. The gear-cutters consequently had to be replaced by cutters manufactured from a sintered material which enabled the requisite cycle times and a four hour cutter life to be achieved. The company had to provide sufficient cutters for 1,000 hours production.

Total time from initial enquiry to installation. This time was not considered to be particularly long bearing in mind the size and complexity of the project, the size and complexity of the truck plant into which the machinery was installed, and the consequent work load on the foreign trade organisation at the pre-contract stage.

Utilisation. The company had received a subsequent enquiry for grinding machines for the production of gear-cutting tools. This enquiry confirmed that the factory was producing its own requirements for gear-cutting tools, although the method used seemed to be old-fashioned with respect to the generation of the tool form. This aspect, together with possibilities of shortcomings in the Soviet-produced tool material is likely to have led to lower levels of tool life and consequent utilisation. In terms of manning and skill levels, however, the Soviet factory was considered to be as efficient as Western European counterparts. The time taken to achieve full production (eighteen months) was considered to be caused by the size and complexity of the complete factory project.

Diffusion. The company did not consider that the USSR would produce identical machinery, since the long term Soviet demand was likely to be small for this type of specialised machine.

Case study no. 5 - a producer of special-purpose machining systems

Technical background. The company which provided the information for this case study is a large machine tool group engaged in the design and manufacture of a wide range of machine tool types. The majority of

Soviet purchases from the company, however, have consisted of conveyor-linked lines of machine tools, specially tooled and equipped with automated loading and unloading devices for the large scale repetitive manufacture of a particular type of automotive component. Most of this case study discussion will consequently focus on an installation in an already established Soviet truck factory, the machines provided by the company being purchased as part of a machine replacement and modernisation programme. Much of the design of this link line, however, was based upon the company's experience in the provision of machinery for the production of a similar component for a passenger car engine, which had been previously installed in a new Soviet factory.

The Soviet truck factory in which the link line was installed was previously producing the particular component by means of specially-tooled and adapted standard machines which were manually loaded. The purchased link-line was consequently more sophisticated in terms of the specialised design of the machines and associated tooling, the automated machine loading and unloading devices, and the conveyor system used to automatically transfer the components from one machining station to the next. In addition, the system was provided with post-process gauging devices located after critical machining stations, based on the company's experience that this type of link-line would operate more efficiently with the requisite amount of post-process gauging devices. Furthermore, at the request of the customer, automatic tool-wear compensation was provided on some operations, using information obtained from the relevant post-process gauging devices.

The company was of the opinion that the Soviet machine tool industry would have been technically capable of producing such a system itself if sufficient priority had been given to the project through the Soviet central planning system. It was considered, however, that difficulties would have been encountered in attempting to integrate various aspects of the project (e.g. the provision of automatic gauging and the provision of conveyor systems), and it was consequently easier for Soviet machine tool users to purchase such packaged expertise from Western companies. Furthermore, when observing newly-installed machinery at the new car engine factory, it was considered that, although Soviet-built transfer machines at that time may have had the same basic technical characteristics as their Western counterparts, they may have been less reliable in conditions of intensive operation. Finally, the load on the Soviet machine tool industry at that time was probably at such a level that there was insufficient available production capacity to produce equipment of that type by the required delivery date.

It was found that the Soviet customer was not always enthusiastic to carry out technical developments to enable the purchased machinery to operate more effectively in his manufacturing system. For example, when problems were encountered in the general quality and machinability of cast raw material, the customer was initially reluctant to acknowledge such shortcomings, perhaps because they were cast in his own factory which was considered to be something of a showpiece in Soviet advanced automobile technology. Similar casting problems had been encountered in the car engine plant, but they had been more readily resolved, perhaps because the foundry was not part of that plant. It was found, however, that the Soviet customer had accounted for various problem areas in production technique during the various stages of equipment proposal - preferences for processing stages and tolerances had been specified, processing methods and tooling configurations had been laid down for certain operations, extra capacity had been requested for some operations, and the actual types of preferred machines had been stated in some cases.

Proposal and contract. The time period between receipt of initial enquiry and submission of final proposal was found to be some four years. The final proposal, although similar in principle, was found to differ quite substantially from those initially submitted. This was partly due to requests from the customer for certain machine and tooling configurations based upon their technical preferences, and also due to certain other technical proposals put forward by the company (e.g. increased amounts of automatic post-process gauging).

The time taken to arrive at a final proposal was considered to be longer than that usually encountered for similar technical complexity, with buyers from other advanced industrialised nations. This was considered to be partly due to the necessity of channelling all communications through the relevant foreign trade organisation, and the fact that a new truck factory was also being built at the same time in the USSR, which may have had priority in the allocation of investment funds and hard currency. Furthermore, since the machines were intended for replacement and modernisation, there may have been less pressure for production capacity than in the case of a new factory.

Acceptance and installation. In view of the size and technical complexity of the project, deliveries were carried out in three discrete stages over a period of two years. In general, there was no lateness in delivery, and some machines were delivered before time. A few delays were sometimes encountered in the carrying out of cutting tests, and subsequent modifications of tooling, if the customer had not always previously provided all of the required technical information. Furthermore, certain modifications were sometimes required to the materials handling equipment when tested under conditions of full loading. The meeting of delivery dates was further complicated for a number of reasons, particularly those related to the co-ordination of some eighteen major sub-contractors.

No contractual dates were given for final acceptance of the equipment on site by the customer.

Total time from initial enquiry to installation. It was the company's opinion that the proposal stage was approximately three times longer than for a project of similar complexity for a West European customer for the reasons discussed in the relevant paragraph above. Installation and commissioning would also take about three times longer because of certain factors on site. These included a lack of provision of appropriate facilities and resources for installation, including equipment and labour of the requisite quantity or quality; the general organisation of the factory in which there would be considerable gaps in skill and expertise between a highly trained engineer in charge of the project and the 'brigade leader' in charge of the appropriate group of production workers; and an unwillingness for personnel to take individual responsibility for acceptance - it was not unusual for twelve of the customer's representatives to be present during acceptance testing of the equipment. Furthermore, although the customer's electricians were usually quite highly trained, mechanical fitters were found to be inadequately equipped and less well trained.

Utilisation. The company did not have much detailed information on the utilisation of the equipment following installation, although it was of the general opinion that the Soviet customer was moving in the right direction for successful assimilation of the equipment. The time requ-ired for successful assimilation was partly due to the general technical complexity of the equipment, but this was also considered to be extended because of the general conditions on site.

The level of manning for production workers was considered to be sligh-tly higher than that which would be encountered in the West, but the level of manning for unsupervised activities such as maintenance was found to be considerably higher. This was considered to be due to the general level of skills and application, the Soviet policy of full empl-oyment tending to lead to a labour surplus, and a general lack of basic labour-saving devices.

Diffusion. The company had no evidence of Soviet machinery incorporating technical features of the company's products. It was considered that over-academic training of Soviet engineers, gaps in skills between eng-ineers and technicians, lack of appropriate incentives to foster techni-cal development and improvements, and shortages in the supplies of re-quired materials, devices and equipment, may tend to hinder the diffus-ion of Western technology throughout the Soviet economy.

Case study no. 6 - a producer of turning machines

Technical background. The company which provided the information for this case study is a designer and manufacturer of horizontal and vertical turning, boring and facing machines. During the late 1960s the company exported thirteen vertical single-spindle machines, specially tooled for repetitive turning and facing operations, and also provided with auto-matic workpiece loading and unloading equipment. The machines were worth some £200,000 and were to be installed in a new passenger car gear-box plant estimated at some £20m in total value. The Soviet buyer was also purchasing technical assistance for this project from a West German producer of gearcutting machines.

The company did not think that Soviet industry produced a machine of this type at that time; although the operations carried out by that machine could possibly be done by other, although less efficient, methods. The company considered that the Soviet machine tool industry had the technical capability to design and manufacture machines of that type, but had ranked the design, development and production of such machines at a low level on its list of priorities. It was also likely that the USSR imported such machinery because the plant in which it was to be installed was to be completely equipped with imported machinery.

The company did not encounter any technical problems on the supply of pre-processed items to its machine - consequently, the company did not consider that the Soviet end-user had to carry out any substantial pro-cess development in order to integrate the company's equipment into his production system.

The purchased machinery was not the subject of any patents for which the Soviet buyer became a licensee.

Proposal and contract. The time between submission of the original technical proposal and agreement on the final technical proposal was found
to take some twelve to eighteen months. The usual time for this task for
similar machines when dealing with Western companies was between three
and six months. The company did not consider this decision-making time
to be especially long, however, when account was taken of the size and
technical complexity of the factory in which the equipment was to be installed - technical decisions concerning the company's products would be
influenced to a certain extent by technical decisions concerning the
plant as a whole.

In addition to these technical factors, the company considered that
decision-making times were also influenced by the normal Soviet bureaucratic nature of doing business, causing communication difficulties between the foreign trade organisation and the end-user, and the presence
of a third-party in the chain of communication - the West German company
which provided technical assistance.

Acceptance and installation. The company found the acceptance procedures
followed by Soviet inspectors to be far more rigorous than those usually
encountered from Western customers, thereby lengthening the usual time
for this stage. For this project, it was found that acceptance took
some ten weeks, whereas similar tasks with a Western company would only
take some four to five weeks. The lengthy acceptance times were considered to be due to concern, on the side of Soviet buyers, that the long
distance between the factory and the seller might give rise to poor after
sales service if problems arose. Furthermore, acceptance times were
sometimes increased by a tendency on behalf of the inspector to select
some of the more rigorous specifications of the West German company which
provided technical assistance. The inspector also selected acceptance
tolerances which were tighter than those specified in the component
process stage drawing, which required substantial modification to the
relevant tooling. Finally, delays in the completion of various stages
on site may also have caused the acceptance of machines to take longer
than normally expected.

The installation time took some twelve man weeks compared with some
three man weeks expected in a Western factory. This increased time was
caused by a number of factors, chiefly related to the overall size of
the complete factory project, and conditions on site. It was sometimes
necessary for other equipment to be positioned before the company's
machines could be installed, and it frequently took a long time to arrange for quite simple modifications to be carried out.

Total time from initial enquiry to final installation. The time elapsing
from initial enquiry to completion of installation was some three times
that which would be expected for a similar sale to a Western company.
The reasons for these lengthy time intervals are given above, but the
chief causes would appear to have been the lengthy pre-contract stage
(one to one and a half years) and installation (three months).

Utilisation. The company had no information on utilisation of its equipment following installation in the customer's factory. It was difficult to express any opinion with regard to levels of manning since, on
the one hand, the Soviet customers were purchasing automatically loaded
and unloaded machines which should have led to a lower-than-normal level
of manning; whilst on the other hand, it was wondered whether the expect-

ed low level of manning would have been achieved in view of the comparative lack of Soviet experience in the operation of automated equipment.

Diffusion. To the best of the company's knowledge, the USSR did not
produce machines similar to those of the company's design. In the opinion of the company, however, this was not due to technical impossibility
(with the possible exception of consistent accuracy) but as a consequence
of resources being invested in other projects deemed to be more important.

Case study no. 7 - a producer of grinding, gearcutting and turning
  machines

Technical background. The company which provided information for this
case study is a designer and builder of high precision thread grinding
machines covering a wide range of applications from very small machines
for precision gauge grinding, through universal and tap grinders, up to
five metre length machines for the grinding of precision ball screws.
In addition, the company produces a range of turning and gearcutting
machines. The company has sold a large number of each of these types of
machine to the USSR in recent years; none of the technical features of
the equipment was covered by patents for which the USSR became a licensee.

To the company's knowledge, the Soviet machine tool industry did not
produce grinding machines and gearcutting machines to the same standards
of quality and precision as those produced by the company, although
Soviet-produced turning machines did not compare particularly unfavour-
ably with the company's products. In the case of grinding machines, the
company did not consider that the Soviet machine tool industry had production capacity to manufacture both the large and small types of thread
grinding machine produced by the company. This was partly because the
demand for these specialised machines would not be sufficiently high to
warrant the required development costs (the company produced approximately 200 machines per year in total: this type of machine was produced
by only two other competitors located in Switzerland and West Germany
respectively); and partly because of the high degrees of skill and long
years of experience required to successfully build these machines and
their associated attachments (e.g. wheel dressing devices). The USSR
had a gearmaking machine production capacity, but the company considered
that the reliability and durability of Soviet-produced machines of this
type were not adequate for high volume production conditions. In general
however, the USSR appeared to be attempting to be self-sufficient in its
production of turning and gearcutting machines, only importing these
items to meet shortfalls in times of high demand (e.g. expansion of automobile factories in the late 1960s and early 1970s).

The company considered the organisation of Soviet engineering production to be adequate for the successful utilisation of its thread grinding
machinery. The workpieces as presented to the machine were satisfactory,
and the techniques of measurement were precise. Furthermore, the factories in which they were installed (probably existing tool or instrument
factories) appeared to have the necessary expertise to install and assimilate these machines. In the case of turning machines (joined together
in a link line), however, it was found that modifications to Soviet-
produced forgings were necessary to achieve the required tool life. This
was achieved as new forging machinery came into operation in the auto-

mobile factory in which the turning machines were installed.

Proposal and contract. The time between submission of first proposal and completion of final technical proposal tended to vary with the novelty of the proposed equipment. In the case when repeat equipment was being offered, inspection of the technical aspects of the proposals might only take a few weeks; but when the machine or its application was new to the Soviet customer, the time scale might be extended to some six months. In certain exceptional cases, machinery may be required in a hurry by the Soviet buyer (perhaps if a factory project was running late) and the pre-contract stage would then be short, perhaps only a few months.

The greatest delays, however, were found to occur during price discussions, and the length of these delays was considered to be greatly influenced by the general level of activity in the machine tool industry outside the USSR. If the Western machine tool industry was generally buoyant to the extent that suppliers could take a firm line on price, then contracts could be closed within six months of submitting the quotation. If the machine tool industry was in a depressed state, however, it was usually found that the Soviet side would continue to argue for lower prices; if the supplier took a firm line against ridiculously low prices, discussions may go on for as long as two years. This compared with the usual nine months in Western markets.

Acceptance and installation. Acceptance of the machines and equipment was usually at the supplier's own works, although buyers reserve the right for final acceptance at the user's factory. In view of the company's long years of successful service to the Soviet market, the purchasing foreign trade organisation was prepared to accept the company's testing documents as approval for acceptance of thread-grinding machines - usually standard machines designed to grind a thread of a certain size and precision. When the company delivered its first large contract, however, it was found that acceptance procedures were unduly long - three to four months instead of the usual three to four weeks in Western markets. A similar lengthy delay was encountered during the acceptance of turning machines, although the major cause of this was the poor machinability of the Soviet-produced forgings leading to low values of tool life. This was finally resolved by the use of forgings from the Western company providing the technical assistance to the factory in which the machinery was to be installed.

The company had never been requested to install or demonstrate the thread grinding machinery in its final locations. In the case of the gearcutting and turning machinery, installation and commissioning was required. In this case, installation took between three and four months whereas approximately three to four weeks would be expected for a similar project in a Western company. These delays were caused by the general organisational conditions in the Soviet factory (e.g. difficulties in obtaining manufacture or modifications, the necessity of waiting until equipment carrying out prior operations had been installed); and the Soviet demands that large numbers of operators be trained by the company's skilled engineers.

Total time from initial enquiry to installation. This was generally found to be longer with Soviet customers than their Western European counterparts, partly because of the size and technical complexity of

orders delivered to Soviet customers, but mainly because of the usual
Soviet procedures of highly contended pricing, and also the delays in
installation previously referred to. In general, proposal, acceptance
and installation would take some eleven months for a Western company –
in the case of a Soviet buyer this cycle would take some two and a half
years.

Utilisation. The company had no information on the utilisation of its
equipment, but considered it to be satisfactory. To the best of its
knowledge, none of the technical features of its equipment had been assimilated into Soviet-produced machinery.

Case study no. 8 – a producer of automotive components

Technical background. The company which provided information for this
case study is a designer and manufacturer of automotive components.
During the late 1960s the USSR purchased a licence for the production of
a clutch unit of the company's design, together with production know-how
including some plant and tooling. The Soviet motor industry, at that
time, was using a coil spring clutch of post-war design but was also
attempting to develop a diaphragm clutch of contemporary design in order
to improve the transmission characteristics of its passenger cars which
it hoped to sell in the international market. Since the Soviet motor
industry did not have the necessary specialised technology it was obliged
to purchase a licence for designs, patents and also manufacturing know-
how. The company's field of technology was one in which there were a
few leading companies some ten years ahead of the remaining component
producers, including those operated by the Soviet motor industry.

The Soviet purchase of manufacturing know-how from the company also
necessitated the purchase of some plant and associated tooling to ensure
that conformity with the required process specification was achieved.
This included assembly and testing equipment to the company's design,
and also tooling and control equipment which could be fitted to more
standardised equipment of Soviet manufacture. The Soviet buyer either
purchased directly from the company, or from a recommended supplier.
The special purpose and assembly equipment was estimated by the company
to account for some 15 per cent of total investment using the normal
Western profile of make-in and purchase. In the Soviet case, however,
this proportion of investment would be lower in view of the tendency to
make-in a large proportion of the items that would be purchased by a
Western manufacturer.

Proposal and contract. The company had been in contact with the Soviet
buyer since the mid-1960s as part of a continuous dialogue on the possibilities of Soviet manufacture of clutches of advanced design. Over a
two year period, the company found it to be necessary to submit a wide
range of proposals from which the Soviet buyer could select a permutation
most advantageous to him from the points of view of cost and transfer of
advanced technology.

The final technical proposal bore little relationship to the original
technical proposal, and the company considered this time interval of
submission of new and modified proposals to be an important learning
period by the Soviet buyer, in addition to the provision of possible
points of leverage in commercial negotiations. The period of technical
discussion was not considered to be much longer than that which would be

generally encountered in discussions with a Western company; the commercial negotiations were found to be extremely protracted, however, taking some eighteen months instead of the usual nine months with a Western company. The total time interval between initial enquiry and contract consequently took some four years compared with some two to two and a half years for a Western company.

Acceptance and installation. The equipment acceptance stage was found to take some three to four times longer than that usually found when exporting similar equipment to a Western company. This was not necessarily due to technical problems - the raw materials provided for acceptance testing were found to be quite satisfactory, but due to the general procedural difficulties encountered (e.g. time required to obtain travel clearances for Soviet inspectors).

The company had no direct involvement in installation and commissioning since this was not part of a contractual condition. From information received, however, it appeared that the installation stage was prolonged as a consequence of the overall size and technical complexity of the complete project - the clutch factory was only one part of a large complex of factories for the production of automotive components. Furthermore, the company had reason to believe that commissioning had been delayed as a consequence of difficulties in the operation and maintenance of the advanced control equipment purchased by the Soviet buyer. These problems were considered to be compounded, however, by the difficulties encountered in the large-quantity production of forgings to the consistent quality required for subsequent monitoring by the control equipment. Difficulties were also encountered with the production of pressings probably as a consequence of the quality of materials supplies.

Total time from enquiry to installation. The time taken at the pre-contract and acceptance stages have been discussed above. The company also considered that a longer time was required to reach normal output levels as a consequence of the difficulties of servicing advanced technology equipment in the remote area chosen for its installation. Furthermore, the company considered Soviet output targets to be extremely ambitious, but considered this to be a characteristic of the international automotive industry in general rather than the USSR in particular.

Utilisation. The company had very little information on the utilisation of the equipment which it supplied to the Soviet buyer. It had reason to believe, however, that the manning levels would be approximately the same as for a Western manufacturer.

Diffusion. The company believes that the USSR has attempted to use the information which it purchased for the design and manufacture of other sizes of clutch, although this has met with little success. Furthermore, much of the purchased information has been made available to other COMECON countries. The company considers that the successful diffusion of technical information throughout the Soviet economy is hindered by the gaps in skills between the shop floor and the product research and development institutes, which in their turn tend to follow rather academic approaches.

COMMENTS AND CONCLUSIONS

It was found that the companies visited during this survey carried out in late 1978 represented a cross-section of British engineering companies engaged in the export marketing of machine tools to the USSR during the late 1960s and the early 1970s. In almost every case, the company was responsible for the design and building of a machining system to repetitively produce one component or a range of components. This usually included the machine tool itself with associated tooling and automated component loading and unloading equipment; component transfer equipment was also provided in some cases. Each company also usually packed the equipment together with associated technical documentation for delivery from a British port, and the majority of companies also subsequently installed and commissioned the plant in a Soviet factory. A licence was sold in one case only, this relating to the product to be manufactured by the equipment (an automotive clutch), rather than any feature of the manufacturing equipment itself.

The executives interviewed in the survey had several years of experience in the Soviet export market and were consequently found to be invaluable information sources on Western technology transfer to the USSR through its foreign trade system. Each of the executives co-operated by providing information on the topics raised in the questionnaire, although not necessarily in an exact fashion for reasons of commercial secrecy or because the information was not always available in the company's files. Nevertheless, it was considered that sufficient information was forthcoming to meet the objectives of the survey, and that the structured interview method served as a useful means of providing a focus for the discussion.

In the majority of cases, the machines were purchased to carry out a specific production task in a particular Soviet factory, usually representing important production capacity to that factory. It was usually felt that Soviet purchasing policy was initiated as a result of shortcomings in Soviet production capacity for machine tools, but once imports were recognised as a necessity, Soviet buyers clearly required advanced production technology at suitable deliveries and prices to achieve required productivity. Most of the executives considered their exported equipment to be technologically superior to comparative Soviet-produced items chiefly in terms of accuracy, reliability, and quality of finish, but it was difficult to relate such parameters to a time period of technological lag.

In most cases the companies commented on the lengthy time intervals between the receipt of an initial enquiry from a Soviet purchaser and final agreement on a technical proposal prior to the signing of a contract. This lengthy time interval could sometimes be explained by the size and technical complexity of the equipment requiring lengthy times for the examination of proposals, although even when allowance was made for this it appeared that in several cases this stage in the purchasing procedure took longer than two years - some three times longer than the normal time with a West European customer. This lengthy time interval was attributed to communication delays within the foreign trade organisation itself and also between the foreign trade organisation and the user factory. The general work load on the two main foreign trade organisations importing machine tools was also high at that time as a consequence of the large number of enquiries and orders being handled by

them. Furthermore, the pre-contract stage was also frequently lengthened by the normal Soviet practice of highly contended commercial negotiation.

It became apparent, therefore, that Soviet bureaucratic procedures
considerably lengthened the initial stage of technology acquisition from
the West. On the other hand, Soviet production engineers had the opportunity to become extremely well-informed on the technical characteristics of the equipment which they wished to buy, through the receipt of
relevant detailed technical information in the proposals from Western
companies, and the opportunity for several technical discussions with
specialists from those companies - a rare and possibly privileged opportunity in the general conditions of restricted and controlled access to
the West. Soviet production engineers should consequently have had
sufficient time and opportunity to make the most rational purchasing
decision in terms of their quality and output requirements. Furthermore,
in the usual Soviet conditions of scarce foreign currency, extended
proposal and negotiation times frequently allowed Soviet buyers to
receive extremely favourable commercial conditions, particularly if
other capital goods markets were generally depressed. The present
author would argue, therefore, that extended purchasing times were probably a small price for Soviet buyers to pay for the technical and commercial conditions thereby gained, although more research may be necessary to clarify this point.

In many cases, the companies had to take great care to meet delivery
dates in view of the generally thorough, and sometimes over-academic,
procedures followed by resident Soviet inspectors; and the lengthy time
intervals to reach decisions if external consultation was required by
the inspector. If great care had not been taken by the company, deliveries could easily have become extended beyond the expected completion
date. From the viewpoint of the Soviet buyer, however, the disadvantages of lateness at this stage may have been a comparatively small
price to pay for the opportunities to gain more technical information
on the operation of the equipment from its designer and manufacturer,
and assurance that the equipment conformed exactly to requirements,
since reference back to the seller after shipping would not have been
easy through the Soviet bureaucracy. Furthermore, since many of the
machines were purchased for installation in new factories or extensions,
conditions on site may have made it expedient to delay the delivery of
certain items of plant.

The most lengthy delays in Soviet assimilation of Western machine tool
technology, however, appeared to occur at the installation and commissioning stages, chiefly as a consequence of Soviet project management
and resource allocation. The installation and operating personnel encountered on site were not always of a very high calibre, and were
seldom sufficiently motivated and adequately equipped by normal Western
standards. Furthermore, the workpieces offered for machine acceptance
and commissioning were frequently of a poorer quality than that anticipated during the original machine proposal stage and testing in the UK,
and other Western imports of casting and forging machinery were frequently necessary to rectify these problems. Such technical and organisational problems frequently caused installation and commissioning
times to be extended to about three times the length of similar installations in advanced Western countries. It is necessary to mention,
however, that differences were noted between the levels of expertise in

project management at different Soviet factory sites - in the case of the latest Soviet new truck factory project it was considered that conditions were generally quite orderly, although this was probably due to the fact that overall responsibility for this factory rested with a Deputy Minister of the Automobile Industry. Furthermore, in some cases, equipment supplied by the companies had been installed with apparent success by the Soviet factories themselves.

It became clear, therefore, that the total time cycle of technology absorption from receipt of enquiry to final commissioning was lengthy in the case of the USSR purchasing Western machinery. It is difficult to put an exact figure on this, but an estimate of between two and three times the expected time span for a factory in an advanced Western nation purchasing a similar item of plant would not appear to be too inaccurate. Many of the causes of these delays would appear to be caused by Soviet bureaucratic inefficiencies and poor project management - this latter aspect appearing to be particularly surprising in view of the fact that a considerable part of the total price of the machine is usually paid for on delivery, and one would consequently expect the customer to be further motivated to rapid installation and commissioning.

The only apparent advantage to the Soviet buyer to mitigate these lengthy time delays would appear to be the opportunity to gain more detailed technical information on his plant purchases, especially by contact with relevant Western engineers at the proposal, build and installation stages when these specialists are usually being stretched to their full technical capabilities. Several Western studies (19) have demonstrated the key role played by the movement of trained individuals in the technology transfer process ('technology on the hoof'), and in view of the general restrictions on Western/Soviet technical contact it is suggested that the opportunities provided by the foreign trade process for frequent and sustained contact by Soviet organisations with trained Western engineers, is the nearest that these organisations could go to this type of technology transfer.

Few of the executives interviewed in this survey had further information on the performance of their equipment following installation, and none had any concrete evidence of the copying of their technical ideas by Soviet engineers; although this latter stage of technology transfer may not occur until the equipment has had a long trial in practice.

Some of the executives interviewed considered Soviet factories to be comparatively overmanned by Western standards, ascribing this to the lower levels of skill and motivation amongst the Soviet workforce, particularly in such unsupervised activities as mechanical maintenance; and the general Soviet policy of full employment.

In conclusion, therefore, these structured interviews with British technical and commercial executives having experience in almost all aspects of technology transfer to the USSR through the foreign trade mechanism, have served as a useful snapshot of various features of Soviet behaviour in the process of absorption of Western technology. Many of these features may appear to be inefficient by Western business economic standards, but by Soviet criteria they may have served as important stages in the successful acquisition of advanced Western manufacturing technology, particularly in the motor industry.

NOTES

(1) See Amann, Berry and Davies (1969), pp.421-424,491-493. The research, development and design competence of these organisations
can be noted from accounts published in the monthly house journal
of the Ministry of the Machine Tool and Tooling Industry - Stanki i
Instrument (an English translation Machines and Tooling is also
published) during that period.

(2) Soviet data for as early as 1964 stated that eight models of machine
tool were produced in quantities greater than 2,000 per year, and
a further 9 models were produced in quantities varying between
1,000 and 2,000 per year (Demchenko (1964), p.273). Furthermore,
at approximately the same time, thirty-two Soviet machine tool
factories produced machines in quantities greater than 1,000 per
year, including eight factories producing more than 3,000 machines
per year (see Chernykh (1965), p.32).

(3) See note (1) above.

(4) Daschenko and Nakhapetyan (1964), pp.140,141, give examples of
special purpose unit machines and transfer lines developed by a
Soviet production engineering establishment (The Scientific-Research
Institute for Production Technology in the Motor Industry) and
factory (ZIL factory, Moscow) within the motor industry.

(5) See Berry and Hill (1977) in which particular attention was paid to
the quality of Soviet milling machines, centre lathes and grinding
machines using Soviet state standards, other Soviet technical publications on machine tools, and interviews with British purchasers of
Soviet-produced machine tools as sources of information. In most
of these cases it was found that the accuracy requirements of Soviet
machine tools, as specified in Soviet state standards, was generally
lower than those adhered to by British producers of similar equipment
during the 1960s, but the situation appeared to have improved somewhat as the USSR has introduced new standards in the 1970s. It is
the present author's opinion, however, that a wider survey of the
quality of more machine types, together with important components
and assemblies, is required to achieve more definite conclusions.

(6) During the 1966-70 and 1971-75 Plans, for example, the Soviet production of cars and trucks increased from some 581,000 in 1965 to
some 869,000 in 1970, and some 1,897,000 in 1975. At the same time,
the Soviet output of machine tools only increased from 186,000 in
1965 to 202,000 in 1970, and 231,000 in 1975. Clearly the rate of
expansion of this important machine tool user industry was far
greater than the rate of increase of Soviet domestic machine tool
capacity, even when the increased values of Soviet machine tool
output during this period are taken into account (712 million roubles
in 1965, 978 million roubles in 1970, and 1,493 million roubles in
1975, all expressed in 1967 factory wholesale prices. (Narodnoe
khozyaistvo SSSR v 1975 godu, pp.259,265)).

(7) Two Soviet texts give a useful account of Soviet designs of this
type of equipment during the mid- to late 1960s namely Yarkov (1965)
and Boitsov (1972); whilst another Soviet text gives a useful
account of the operation of these machines in Soviet industrial
practice (see Volchkevich (1969)). For a further account of the
operation of these types of equipment in practice in the mid 1970s
see Machines and Tooling (1972), No. 2, pp.5-10, and Machines and
Tooling (1975), No. 8, passim.

(8) See Berry and Cooper (1977) for a description of Soviet design and
technology in the field of numerically controlled machine tools.

(9) See, for example, a survey carried out by the Research and Technical Committee of the Institution for Production Engineers which included the question of imports of certain machine tools into the UK, (Institution of Production Engineers (1968)).

(10) See, for example, Hanson (1976 & 1978) and the Chase World Information Corporation (1977), pp.133-174.

(11) See Hanson and Hill (1979).

(12)

	1970	1971	1972	1973	1974	1975
Western exports of machine tools to USSR	152.8	101.8	212.7	306.8	372.4	451.3
Western exports of machine tools to world	1601.3	1701.2	1824.9	2385.6	3123.1	3708.9
Percentage of Western total exports delivered to USSR	10%	6%	12%	13%	12%	12%

All export data in $ millions FOB, abstracted from annual publications of *Bulletin of Statistics on World Trade in Engineering Products* (ECE, NY), UK, France, Italy, FRG, USA, Japan grouped together into 'Western exporters'.

(13)

	1965	1966	1970	1971	1974	1975
UK exports of machine tools to USSR	4.6	3.9	21.4	14.9	8.7	12.9
Western exports of machine tools to USSR	15	12.7	152.8	101.8	372.4	451.3
British share of Western machine tool exports to USSR	30%	31%	14%	15%	2%	3%

All export data in $ millions FOB, abstracted from annual publications of *Bulletin of Statistics on World Trade in Engineering Products* (ECE, NY), UK, France, Italy, FRG, USA, Japan grouped together into 'Western exporters'.

(14) See Hill (1979).

(15) i.e. the company carried out machine specification, design, construction, installation and commissioning in accordance with accuracy requirements from the customer's component drawing, the customer's output requirements, and other general technical conditions.

(16) The Soviet state standard for acceptance testing of multi-spindle automatic turning machines gives a series of tests and tolerances for 'normal precision' (N class) machines and 'improved precision' (P class) machines. In general, tolerances quoted for P class machines are 0.6 of those quoted for N class machines. (See Hill 1973b).

(17) The drafting of machine tool state standards for approval by the State Committee of Standards of the USSR is the responsibility of the leading Soviet research institute for machine tools (Experimental Scientific Research Institute for Metalcutting Machines (ENIMS) . It is likely that consultation would take place between ENIMS and the leading producers of the relevant machines (the Gor'kii Factory of Automatic Machines, Kiev, in this case), (see Hill (1970), pp.48-57 and Hill (1973a)).

(18) As mentioned previously, the company had also exported approximately 100 single-spindle machines in the late 1960s. It was the company's opinion that no significant technical lag was apparent in Soviet

design and manufacture of these types of machines, but purchases
were required because of shortages in Soviet domestic productive
capacity.
(19) See, for example, Langrish (1972), pp.42-49.

4 East-West industrial co-operation and technology transfer — case studies of British firms in the engineering sector

INTRODUCTION

For the purposes of the study described in this chapter, an industrial co-operation agreement was considered to contain the following elements:

(a) a transfer of technology from the Western partner of the agreement to the Eastern partner for the design, development and/or manufacture of a particular finished product. The technology transfer may also have included the sale of appropriate licences, manufacturing equipment, or components; and the training of personnel;
(b) the time interval of the industrial co-operation agreement was usually longer than that associated with a one-off sale;
(c) if appropriate, payment for the technology transferred was made partly in items related to the finished product.

It can be seen, consequently, that the study was concerned only with aspects of what has been defined by McMillan (1) as 'inter-firm industrial co-operation'; namely industrial co-operation between two parties having the legal competence to carry out commercial activities. This definition consequently excluded co-operation activities between official inter-governmental groups set up under the auspices of bilateral co-operation agreements, and 'technical co-operation agreements' or 'framework agreements' between Western companies and those socialist governmental bodies responsible for the development and diffusion of technology (e.g. State Committees for Science and Technology). These latter activities are discussed in Chapter 7 below.

Several writers have discussed the supposed advantages of East-West inter-firm industrial co-operation, to both the Western and Eastern partners. The considered advantages to Western firms in the furtherance of industrial co-operation with Eastern European foreign trade organisations can be summarised from Wilczynski (2) and Richman (3) as follows:

(a) profitability from marketing, technical and management fees, supply and service contracts, and other arrangements surrounding such ventures, particularly where Eastern European currency limitations may have prevented business from taking place;
(b) the gaining of access to new markets in more of the Eastern European countries as a consequence of industrial co-operation with one of

them; furthermore, entry may be made into new markets in third countries with whom the Eastern European country has had a successful trading relationship;

(c) the socialist countries of Eastern European represent valuable reservoirs of certain raw material and power resources;

(d) the labour resources of the socialist countries of Eastern Europe are relatively abundant, reliable and inexpensive. Consequently industrial co-operation with Eastern Europe may provide access to cheaper sources of supply of components and finished products;

(e) industrial co-operation provides opportunities for specialisation and economies of scale, particularly in those industries requiring large investment outlays on research and development, and capital equipment. Production co-operation with a socialist enterprise may enable a Western company to reduce its overhead costs and draw on its Eastern European partner's capacity;

(f) the Western company may gain easier access to Eastern European technological know-how in an area in which this is relatively well-advanced.

It appears, however, that the advantages of industrial co-operation to the socialist countries are more 'varied and compelling' than to their capitalist counterparts, to quote Wilcyznski (4), who lists these microeconomic and macroeconomic advantages as follows:

(a) the possibility of equipping the co-operating socialist enterprises with modern Western machinery and equipment without direct expenditure of hard foreign exchange;

(b) the direct assimilation of Western technological processes;

(c) the application of modern Western management techniques and know-how, which cannot be acquired otherwise;

(d) economies of scale consequent upon specialisation and production for a wider market;

(e) improvement in the quality of production as the socialist enterprise has to meet the demanding requirements of the Western partner operating in competitive buyers' markets;

(f) improvement of supplies to the socialist domestic market with regard to goods previously not produced, or in short supply;

(g) circumvention of the Western countries' import restrictions as co-operative imports are often admitted under special concessions;

(h) establishment of a foothold in, or the extension of, markets in capitalist countries using the Western partner's marketing channels and expertise;

(i) improvement in the balance of payments consequent upon the saving of foreign exchange on co-operative supplies of equipment by the Western partner, and expanded sales of associated products in capitalist markets;

(j) a disciplining factor in general, prodding the socialist enterprise to more radical and continuous improvements in its methods of production, management, marketing and overall efficiency (5).

The considered advantages of industrial co-operation to Western companies, as listed above, may have to be tempered, however, with certain possible shortcomings frequently encountered when embarking on these types of agreement. The disadvantages can be summarised from Hill (6), and Levcik and Stankovsky (7) as follows:

(a) high initial marketing costs to secure the agreement, since business discussions with Eastern European foreign trade organisations are usually lengthy;

(b) possible problems in the quality and delivery of goods received in payment from the Eastern European partner, as enterprises in the socialist countries do not always have the required awareness of quality and delivery requirements encountered in Western markets;

(c) it may be difficult for Western companies to accurately predict future market and cost conditions over the timespan of the co-operation agreement.

Hovanyi and Ellis (8) also sound a note of caution with regard to the effects of industrial co-operation on Eastern European industry, and list the possible risks of co-operation with Western companies as follows:

(a) a possible decline in the East European partner's own research and development activity, and possible over-dependence on the technology of the Western partner;

(b) uncertainty regarding the adaptation of the Western technology to Eastern European conditions;

(c) problems in the Western economies (e.g. inflation) tending to be transferred to the Eastern European partners;

(d) the Eastern European partner may be excluded from certain markets.

Furthermore, Levcik and Stankovsky (9) highlight that some 'advantages' of industrial co-operation (e.g. provision of improved products for export) to the Eastern European foreign trade organisation (which is usually the commercial partner to the East-West co-operation agreement), may consequently lead to manufacturing problems in the East European enterprises responsible for the actual fulfilment of the product output and quality requirements of the agreement. Problems may be encountered in assimilating the technology which is being transferred through the agreement itself, and the solution of these may also draw resources from the already existing and successful production systems in the enterprise. These aspects of technical innovation are not peculiar to Eastern Europe alone, however, but occur to a greater or lesser extent in all economies.

In view of the accredited advantages and disadvantages of East-West industrial co-operation agreements, and the estimated four-fold growth of the quantity of this type of business arrangement from the mid 1960s to the mid 1970s (see Chapter 2 above), it was decided to supplement the small number of already published accounts of the experiences of British companies in this field of business activity (10) with case study material from a further sample of British firms. The sample of firms were selected from two published sources, namely: Wilczynski (11) and Dragilev et al. (12), and information from two academic specialists in the area of East-West trade (Dr P. Hanson, Centre for Russian and East European Studies, Birmingham University and Professor C.H. McMillan, Institute of Soviet and East European Studies, Carleton University, Ottawa). The companies were initially approached by the author through an introductory letter, and those firms which subsequently agreed to participate in the study were forwarded a set of questions (see Appendix B) relating to the following:-

the number and type of industrial co-operation arrangements;
motivation for the Eastern European partners to enter into industrial co-operation arrangements;

motivation for the British company to enter into industrial
co-operation agreements;
financial aspects of the industrial co-operation arrangements;
general marketing methods for industrial co-operation arrangements.

These topics were selected to coincide with those previously chosen
by McMillan in a postal survey of industrial co-operation practices (13)
and used by the present author for structured interviews with British
companies engaged in Anglo-Soviet industrial co-operation (14). Six of
the companies were subsequently visited by the author during 1979, and
a further four companies were visited during 1981. Approximately half
a working day was spent with one of the company's senior executives
discussing the company's policy, procedures and experiences in industrial co-operation with the socialist countries of Eastern Europe, using
the questionnaire which had been previously sent as a basis for that
discussion.

The ten companies which co-operated in the survey, out of a sample of
fourteen companies originally approached, were engaged in the following
fields of industrial activity:

the design and manufacture of cranes;
the design and manufacture of mining equipment;
the design and manufacture of automotive electrical equipment;
the design and manufacture of automotive suspension equipment;
the design and manufacture of transformers;
the design and manufacture of transport equipment;
the design and manufacture of computer equipment;
nuclear engineering;
the manufacture of agricultural equipment;
the design and manufacture of aero-engines.

This sample of engineering companies were considered to provide useful
information on the practice of industrial co-operation, since agreements
for the mechanical engineering, transport equipment, electrical engineering and electronics industries were found to account for between 50
per cent and 60 per cent of the total population of industrial co-operation agreements surveyed by the ECE (15) in 1976 and 1978. A standard
format has been followed in the composition of the case studies, namely
a brief description of the product range of the company and its previous
experience in the East European market; a description of the co-operation agreement and its financial arrangements, and the advantages and
disadvantages of the agreement to both sides, as viewed by the company.

CASE STUDIES OF BRITISH ENGINEERING COMPANIES ENGAGED IN INDUSTRIAL
CO-OPERATION WITH EASTERN EUROPEAN FOREIGN TRADE ORGANISATIONS

Case study no. 1 - a designer and manufacturer of mobile cranes

The product range of this company includes diesel-powered hydraulic
mobile cranes, rough terrain cranes, crawler-tracked cranes, diesel-
powered mechanical mobile cranes, truck-mounted cranes and high quay
cranes. The company's sales to Eastern Europe (approximately 5 cranes
per year, including Yugoslavia) account for 8 per cent to 10 per cent
of its exports, which in their turn usually account for 70 per cent to
75 per cent of total trade turnover. The company follows the usual

procedures for marketing to Eastern Europe, namely frequent visits to
foreign trade organisations, and promotion by consignment of cranes to
exhibitions. The whole of the region is the responsibility of one
marketing executive, and Yugoslavia and Hungary have been found to be
the largest single crane purchasers in that region.

In 1966, the company signed one of the first East-West industrial co-
operation agreements, with a Polish foreign trade organisation responsible for the import and export of cranes and other handling equipment;
cranes had previously been sold to Poland, but not in any great quantity.
The agreement lapsed in 1973 and no new agreement has been subsequently
signed, although trade in some of the items covered by the arrangement
was continued. The industrial co-operation agreement was administered
separately from marketing in the company - since the majority of aspects
in the agreement related to production, it was administered by the
company's director responsible for the manufacturing function. The
various topics covered by the agreement are outlined below:

(a) the British company provided full design and manufacturing documen-
    tation for the production of a particular model of crane of thirty
    tons lifting capacity, available in wheeled and tracked variants;
(b) the Polish foreign trade organisation delivered a programme of
    twenty three items to the company including four sorts of chassis,
    gearboxes, slew rings and travel units to the company's technical
    specification, for subsequent assembly into various end-items in the
    company's product range. The Polish manufactured items were de-
    livered to the company in mutually agreed quantities and at advan-
    tageous prices - far cheaper than those quoted by UK suppliers;
(c) the company supplied kits of items for use in the crane for which
    the Polish foreign trade organisation had been provided with design
    and manufacturing documentation, but which Polish enterprises were
    unable to deliver (e.g. oil seals, bearings, clutches etc.). These
    items were supplied at mutually agreed prices. Some free-issue
    components were also supplied by the company for assembly into the
    items delivered to it by the Polish foreign trade organisation
    (e.g. rams and bearings for assembly into chassis);
(d) the co-operation agreement was extended to include the sub-contracted
    production to Poland of twelve complete cranes of a modified version
    of the type covered by the agreement. This occurred at a time when
    the capacity of the British company was overloaded;
(e) the agreement also specified countries in which each partner had
    exclusive and non-exclusive marketing rights for the crane covered
    by the co-operation agreement. The Polish foreign trade organisa-
    tion had exclusive rights in the socialist countries of Asia and
    Eastern Europe, and India; while the British company retained ex-
    clusive marketing rights in the rest of the world. The Polish
    partner was also allowed to sell in some other Middle Eastern,
    African and Asian countries, and Yugoslavia, provided that permission
    was first obtained from the British company.

In summary, therefore, it can be seen that the co-operation agreement
included a range of activities, namely:

    a licensing agreement;
    supply of parts to the East European partner;
    provision of parts to the British company's specifications for
    inclusion in its final product;

provision by the East European partner of products produced to
the British company's specification for subsequent marketing
by it;
exercise of quality control by the British company;
an agreement for marketing and servicing in specified
geographical areas.

The quantity of trade covered by the agreement amounted to some £1½
million in 1972, but had fallen to about £¼ million by 1976. The
financial aspects of the agreement can be summarised as payment in
components and assemblies at relatively cheaper prices by the Polish
side, for technology supplied by the British company. The Polish partner still requires to purchase certain items, and attempts are made to
balance the value of bilateral trade flows between the two sides.
Meetings are held annually to fix prices, usually on the basis of macro-
economic conditions prevailing in both countries at the time, and a
common price adjustment factor is then applied right across the range of
relevant items.

The company considered that the major reason for the Eastern European
partner entering the co-operation agreement was the rapid absorption of
design and production know-how to meet its expanding market needs (16).
Furthermore, the outlay of foreign exchange was minimised, since no
immediate currency payment was made for the acquisition of the technological know-how, and attempts are made annually to balance the values
of the bilateral trade in components and sub-assemblies. The company
itself entered into the industrial co-operation agreement for reasons
of cost savings in the production of certain types of sub-assemblies
and components, due to cheaper availabilities of Polish skilled labour,
capacity and capital resources.

The company, however, met with a number of problems in the implementation of the co-operation agreement. These can be summarised as follows:

(a) a tendency for the Polish side to require firm orders a longer time
    in advance than is necessary by British suppliers. This could
    sometimes lead to supplies arriving in larger quantities than re-
    quired (i.e. 'lumpy' supply). Furthermore, delivery lead times for
    certain items could be very long. It would consequently appear
    that the cheaper Polish products were purchased at the expense of
    flexibility in delivery;
(b) a tendency for the Polish side to sometimes cancel letters of credit
    for those items to be supplied by the British company. This could
    lead to the British company being left with comparatively high
    stocks due to postponed delivery;
(c) the necessity of informing the Department of Industry, at frequent
    intervals, that the Polish-produced items were still available at
    a substantial cost advantage. This consequently entailed having to
    unnecessarily contact a number of British companies to obtain
    competitive quotations for supplies of relevant items;
(d) the final and major problem has been caused by an action by the
    Polish side, which was viewed by the British company as not in
    accordance with the spirit of the agreement. The company had put a
    great deal of effort into the development of a certain Middle
    Eastern market, which was its exclusive market under the terms of
    the agreement. Fifty-six cranes had been sold, operators had been
    trained, and after-sales service had been established. In mid-1976,

the company bid for a further 140 cranes, worth approximately £20 million; the Polish side, however, bid for the same contract, quoting cranes which were the subject of the co-operation agreement, at prices equivalent to approximately 50 per cent of those quoted by the British company. The Polish side was awarded the contract, which subsequently caused an increase in demand on Polish production capacity leading to an unavailability of chassis for the British company. The company consequently attempted to make itself independent from Polish sources of supply for those products previously covered by the co-operation agreement. As a result of this experience, it was the company's view that co-operation agreements in Eastern Europe were best restricted to those product areas in which there is a rapidly changing technology, to prevent unwanted market competition.

Case study no. 2 - a designer and manufacturer of mining equipment

The powered roof supports designed and manufactured by this company can be considered to be made up of two distinct elements:

(a) steel fabrications which act as structural and foundation members;
(b) hydraulic power and control systems which locate the load bearing steel fabrications in the required position.

Many of the fabrications and hydraulic units produced by the company have been standardised, but each contract usually requires a specific configuration of these standardised elements, together with some specially designed structural members and hydraulic supports.

It is usual for the company to quote against an enquiry for the provision of equipment for a particular mine, and this method of trade is also normally followed for the Eastern European socialist states with which the company deals (Hungary, Romania and USSR), although on a comparatively small scale (between 0 per cent to 5 per cent of total exports annually; exports usually account for about 15 per cent of total annual turnover). The contracts usually include design, production and delivery of the equipment against the customer's specification, followed by installation in an end-user's coal mine. Since the late 1960s, however, the company has signed three separate contracts (in 1969, 1971 and 1973) with the Hungarian foreign trade enterprise responsible for import and export of heavy engineering equipment. These contracts for equipment for three separate Hungarian coal mines contained the following elements:

(a) all technical proposal and design work was the responsibility of the British company;
(b) production of the hydraulic and electrical equipment, and other items which could not be manufactured in Hungary, was carried out by the British company and purchased by the Hungarian foreign trade organisation using letters of credit;
(c) production of the fabrications was carried out by Hungarian enterprises, usually attached to the mining 'trust' in which the equipment was to be installed. These fabrications were produced to the company's drawings provided free of charge. The company also provided free technological know-how where necessary (e.g. training of welders to pass National Coal Board proficiency tests and exercise of quality control). The value of these fabrications

accounted for some 50 per cent to 65 per cent of the total value of
the equipment;

(d) the British company was responsible for supervising the assembly of
the equipment in the Hungarian engineering works, and subsequent
installation in the Hungarian coal mine. The company considered
that Hungarian installation technique was up to Western standards.

The contract consequently contained the following elements of an industrial co-operation agreement:

> training of Eastern European personnel;
> provision of technical assistance (know-how);
> supply of parts and components to the Eastern European partner;
> provision, by the Eastern European partner, of parts and components
> to the British company's specification for assembly in its product;
> exercise of quality control by the British company, (chiefly in
> welding, assembly and installation).

As a result of experience gained from these contacts, the company also
subsequently sub-contracted some fabrication work to Hungary, when their
own capacity was over-committed.

The British company considered that there were several technical and
economic reasons for the Hungarian foreign trade organisation entering
into the industrial co-operation agreement. These reasons can be listed
as follows:

(a) the outlay of a lower level of foreign exchange, since it was only
necessary to pay for the hydraulic items in hard currency, instead
of the complete roof support units. Consequently, the outlay of
foreign currency was probably halved for the same quantity of the
company's designed roof supports installed in Hungarian mines.
These units, in their turn, were considered by the company to be
technically superior to other similar items available on the
Hungarian market, in terms of improved reliability and lower maintenance costs;

(b) transport costs would be substantially reduced, since the fabricated
items account for the bulk of the weight of the completed roof
support unit;

(c) the UK company had substantial cost advantages in production of the
requisite hydraulic equipment because of the availability of skilled
labour, equipment and production expertise. Such cost advantages
were not apparent, however, in the field of fabrication;

(d) the Hungarian enterprises were able to quickly absorb some of the
general technological know-how required for the complete design and
production of hydraulically activated roof support units. In consequence, these enterprises had subsequently produced a hydraulically operated support unit, although of a different design (a 'shield
unit') in place of a 'chock unit'). It was still necessary for
hydraulic items to be imported, however. Furthermore, the Hungarian
enterprise had received spin-offs from training of welders and the
implementation of quality control procedures. The company found the
technical capacity for fabrication of Hungarian enterprises to be
completely acceptable;

(e) the equipment secured under the co-operation agreement was required
for the development of a new coalfield (17), in which there was to
be one new installation per year over a time interval of six years.

The quantities of production envisaged for this equipment did not justify setting up of production facilities for the complete equipment by the Hungarian partner, but fabrication capacity was already available.

The company itself had entered into the industrial co-operation agreement in order to protect its existing market in Hungary. Although the total exported sales volume was decreased as a result of the Eastern partner purchasing only hydraulic items, it still enabled the company to achieve profitable sales of these items in a market which may have completely disappeared. Furthermore, it enabled the company to have an alternative source of supply of fabrications, although Hungarian prices were not found to be particularly competitive, probably as a consequence of transport costs.

In conclusion, the company appeared to be well satisfied with the business arrangement described in this case study, since it enabled them to remain in the Hungarian market in a profitable manner. Furthermore, it provided adequate, although not cheap, fabrication capacity when the company's resources were overloaded. In general, the company found an industrial co-operation type of arrangement to be better than barter trading proposed by its Romanian purchaser, since a third party was not required. The company remarked on a usual feature of trade with Eastern Europe which also occurred in these business arrangements, namely the lengthy pre-contract negotiations covering specifications, delivery and price.

Case study no. 3 - an automotive and aerospace engineering company

The company which provided information for this case study is a large British company with manufacturing, marketing and service facilities in both UK and overseas. The company is organised on a divisional basis, covering the development, production and marketing of electrical, braking and fuel injection equipment for the automotive and aerospace industries, and control systems for manufacturing equipment. The company has tended to offer standard products in the Eastern European market, and to respond to enquiries. Each individual operating company within the divisionalised overall structure has carried out its own marketing, and establishment of trade and co-operation agreements, usually having one manager responsible for Eastern Europe. The company also has a central division responsible for co-ordinating trading activities with Eastern Europe, which has included the setting up of an Eastern European Committee comprising representatives from each operating company.

Promotion in the Eastern European market has been achieved by participation in a general exhibition at least once per year (usually the Poznan Fair), and specialised exhibitions where appropriate (e.g. the Agricultural Equipment Exhibition in Moscow, the Engineering Exhibition in Brno, and general engineering exhibitions in Bucharest and Plovdiv). Furthermore, visits to markets are made at least once every year, together with frequent visits to commercial attachés. The company is also active in the Czechoslovakian and GDR Committees of the London Chamber of Commerce, the British-Soviet Chamber of Commerce, and the East European Committee of the Birmingham Chamber of Commerce. The company's sales to Eastern Europe account for approximately 1 per cent of turnover.

The company has entered into one industrial co-operation arrangement in Eastern Europe, with the foreign trade organisation for the Polish automotive industry. The ten-year agreement was signed during the early part of 1978, and covers the sale of a licence and the provision of relevant design and technical information for a tractor alternator, with the possibility of subsequent variants for use in cars and light aircraft.

The co-operation agreement stemmed from previous trade arrangements with Poland, particularly the licensing, implementation and sale of plant for the manufacture of fuel injection equipment used in a British tractor engine being made under licence in Poland. In that case, however, the licensing and implementation arrangement was concluded with the Polish foreign trade enterprise responsible for the export and import of agricultural machinery; and the sale of plant was arranged with the Polish foreign trade enterprise responsible for the export and import of machine tools. In addition the company has sold licences, implementation and plant to the USSR for the production of brake shoes and high tension coils, and has also purchased light bulbs from Hungary.

The co-operation agreement with the Polish foreign trade organisation includes the following activities:

a licensing agreement;

provision of relevant technical documentation;

training of Eastern European personnel in the area of 'assimilation' of technical documentation;

initial supply of critical parts to the Eastern European partner, although these supplies will diminish as the partner's production competence increases;

a marketing agreement whereby the Polish partner is restricted to the sale of the alternators to COMECON countries and Italy, although the company has no control over the sale of alternators in finished tractors;

the provision of updated technical information to the Eastern European partner during the time of the agreement. The partner is also to observe confidentiality for a period of five years after the agreement has lapsed;

provision by the Eastern European partner of products or components covered by the licensing agreement, for marketing by the British company;

provision by the Eastern European partner of other components for incorporation into other items in the company's product line.

In addition, there is a possibility of the sale by the British company of certain items of plant associated with the manufacture of the licensed product.

The company was paid an agreed sum in sterling for the licence and associated know-how, following the transfer of the relevant technical documentation to the Polish partner. This sum is subsequently to be spent by the British company on products manufactured by factories responsible to the Polish Ministry of Machine Industry, over the time-span of the co-operation agreement. This sum is to be divided in such a manner that 25 per cent of the value is to be spent on components and finished products covered by the licence and the remaining 75 per cent on other engineering products which can be assembled into the company's finished products. The company considers that these items will be useful sources of supply to supplement UK manufacture.

The company considered that the major factor motivating the Eastern European partner to enter into the industrial co-operation agreement was the potential for saving foreign currency in meeting the high continuous demand for alternators in Poland, occasioned by the increases in domestic output of tractors and cars (18). Furthermore, there was a possibility of earning income from exports, including hard currency from some Western markets. In addition to these direct economic benefits, the Eastern European partner obtained a speedy transfer of technological know-how relevant to the technology of contemporary alternators, and also technical spin-offs since the original design of tractor alternator was modified for use in a car, and subsequently for possible use in a light aircraft. The company questioned, however, whether the alternators could be produced more cheaply in Poland than purchased from the UK, since the Polish production quantities would be far smaller.

The British company entered the co-operation agreement in order to protect its market opportunities for licensing in Poland.

With regard to the technical background of technology absorption by the Polish partner, the company provided technical information on its current design of standard alternator with regulator. This product had been manufactured by the company for some ten years, with a series of improvements being carried out over this period. The Polish partner has recently built a large new factory for the production of DC motors, and the company considered that this factory was quite capable of manufacturing the licensed alternator to the required specification and quality.

The company considered that the length of time prior to the signing of the agreement was longer than for licences in Western markets, partly as a result of the more bureaucratic method of management and financing in the Eastern European market - an agreement was finally signed with the Polish partner after some two and a half years of discussions. The company had kept to its schedule for the provision of technical documentation, and the Polish partner had been similarly punctual in its payments.

In terms of technology diffusion, the company considered that a Western licensee might have carried out more rapid updating and improvement, whereas the East European partner was more concerned with the acquisition of 'chunks' of specific contemporary technology for a rapidly expanding industry; attempting to ensure that expenditure of foreign currency was likely to be covered by future sales to hard currency markets.

Case study no. 4 - a designer and manufacturer of power transformers

The company which provided information for this case study is a designer and producer of medium to large capacity power transformers. It has sold its products to Poland and Czechoslovakia, although sales to Eastern Europe have been comparatively small (approximately 2 per cent of total annual turnover), since most of the region is self-sufficient in manufacturing capacity of these products for strategic reasons. The company has sold into Eastern Europe through an agency, but following a merger with a large electrical engineering group in 1968, the company has also been able to draw on the services of the group's corporate specialist for the region.

The company has signed one co-operation agreement with an Eastern European business partner, namely the Hungarian foreign trade organisation responsible for the import and export of electrical transmission equipment. The agreement was originally signed in 1967, for a period of ten years, and was consequently extended for a further five year period up to 1982.

The co-operation agreement relates to the provision of technical know-how in the design and manufacture of power transformers of 3MVA capacity and greater than 200 kV. The agreement involved the sale of a licence, and the provision of associated design documentation and manufacturing advice. Furthermore, updating was to be achieved through the exchange of scientific and technical information which includes bi-annual visits of Hungarian engineers to the British company, which usually last for one week. The Hungarian partner was allowed to sell the licensed product in any area of the world with the exception of UK, the industrially developed Commonwealth countries, South Africa and USA. The British firm, in its turn, was not allowed to provide a licence for the same technology to any other organisation in Hungary, nor to sell any product in Hungary which incorporated technical know-how obtained from the Hungarian partner.

The company receives annual royalty payments in sterling for the use of its licence. There is a minimum annual payment agreed between the two sides, with total royalty payments based on the nominal capacity (in MVA) of Hungarian transformers produced over the year, using the British company's licence. The company has experienced no payment problems whatsoever.

The company considered that the major reason for the Eastern European partner entering the co-operation agreement was the rapid assimilation of technical know-how to meet its expanding markets, both at home and overseas, in certain developing Arabian and African states (19). The Hungarian partner should consequently be able to secure reasonably high export earnings from overseas markets, whilst still meeting domestic requirements with a comparatively small outlay of foreign currency.

The factory which was to utilise the know-how has a long history of electrical engineering know-how, being established since 1844. It is claimed that one of its engineers was responsible for the invention of the power transformer in 1870, but for various reasons, the enterprise had not remained in the vanguard of technical innovation in transformers, although the co-operation agreement with the company has enabled it to quickly raise the level of its technology in that product area. The company found the level of technical knowledge to be comparatively high in the Hungarian factory, as evidenced by the fact that the majority of thermal problems associated with this technology had been resolved leaving only electrical problems.

The company entered the co-operation agreement for reasons which relate to its general marketing policy. The company considers that there are two basically different markets for its products, namely a partly protected home market, and a completely competitive export market. The company's opportunity in this export market has been decreasing for three main reasons, namely:

(a) the growth in electrical load has tended to be low;

(b) each industrial nation has preferred to establish its own power generation and distribution industries, partly for strategic reasons;

(c) Japanese price competition has become almost impossible to counter, since all of the five major Japanese competitors are multi-product firms who can subsidise their transformer prices with profits from the sale of other products.

Since the company is faced with a shrinking market, although it is still a world leader in electrical engineering, it has decided to exploit the market for technological know-how amongst the domestic producers in overseas countries. The sale of licences and associated know-how to Hungary falls within this general marketing policy.

Case study no. 5 - a manufacturer of transport equipment and associated components

The company which provided information for this case study is a designer and manufacturer of transport equipment and automotive components. In addition, it frequently designs and manufactures production equipment for use in its own industry. The company has many years of experience of export marketing to Eastern Europe, maintaining a specialised office for the furtherance of East-West trade on a company-wide basis.

The company has signed two co-operation agreements with Eastern European organisations: the first for the co-operative development of a special purpose casting machine with a Czechoslovakian enterprise, signed in the late 1960s and the second for the production under licence of container handling equipment with a Bulgarian foreign trade organisation, signed in the early 1970s. The Bulgarian agreement was for five years, but no time limit was set for the Czechoslovakian agreement since no accurate time limit could be placed upon the technical development required. The Czechoslovakian agreement stemmed from earlier business contact; the Bulgarian agreement followed as a consequence of frequent contact, although no previous business had been conducted between the two sides. Both agreements have been allowed to lapse.

The Czechoslovakian agreement involved the co-ordination of research and development, exchange of scientific and technical information, and the provision of technical assistance and know-how. It was intended to extend co-operation to a marketing agreement once the technical co-operation was successfully completed, but this option for commercial exploitation of the product was not taken up by either side. The Bulgarian agreement, on the other hand, was to cover:

a licensing agreement;
provision of technical assistance and know-how;
training of Eastern European personnel;
supply of parts and components to the Eastern European partner;
provision of parts and components by the Eastern European partner for incorporation into the final product.

In the case of the Czechoslovakian agreement, the bulk of the costs were current development expenses, which were met independently by each side as and when they were incurred over the five year technical development period. The Bulgarian partner, on the other hand, paid an initial lump sum in sterling for the licence and associated know-how for the container handling equipment, and also purchased certain critical parts and components from the British partner (e.g. hydraulic components). The British company, in its turn, was obliged to buy back certain Bulgarian-made items for use in container handling equipment, and other materials handling equipment for distribution through its agency network. These purchases accounted for a comparatively small percentage of the value of the licence.

Both of the Eastern European organisations entered into the co-operation agreements to improve the export potential of their respective industries. The Czechoslovakian partner appeared to have no particular market in mind, but the Bulgarian partner intended to export (20) the product manufactured under licence to the large Soviet market. Furthermore, the Czechoslovakian partner viewed the agreement as a means to improve the level of innovation in domestic casting technique, saving costs due to the UK company's advantage in technological expertise.

The British company entered into the agreements to expand, protect and diversify its market position and opportunities in Eastern Europe, as part of its general marketing philosophy of reacting positively to market opportunities as they arise in Eastern Europe, and also of creating new market opportunities as industrial needs are detected in that region. Furthermore, industrial co-operation enabled closer working relationships to be established and maintained, which is an important objective when operating in the East European market.

In the agreement with the Czechoslovakian organisation, there was an added incentive to modify a new product, and to subsequently market the modified variant. The company had developed a multi-station moulding machine for the mass production of a range of automotive components up to the size of the average brake drum, and had subsequently identified Eastern Europe as a market for this product in view of the expansion and technological improvement of the COMECON motor industries. The product had been promoted in the region, but especially in Czechoslovakia where it was considered that a particular need existed for this type of machine (21). The Czechoslovakian partner was interested in developing the machine jointly with the company, in order that the equipment be capable of casting a ring-shaped component which had a comparatively short casting time. This possibility of shared development costs was attractive to the company.

The company considered that both agreements created favourable conditions for closer working relationships with their Eastern European partners, giving consequent benefits in a market which tends to operate in a bureaucratic fashion. Furthermore, both agreements were conducted smoothly from the viewpoint of payment.

The agreement with the Czechoslovakian enterprise was not extended into a marketing arrangement for two major reasons, namely that the Czechoslovakian partner appeared to be more interested in technical development for its domestic needs, and also that the British company tends to view itself chiefly as a seller of products and components, not specialised production equipment for their manufacture.

The agreement on the production and marketing of container handling equipment encountered certain difficulties which the company would attempt to remedy in any subsequent co-operation agreement, by the following means:

(a) more comprehensive assistance during the technology transfer process, in order that the design and development expertise of the East European partner be fully translated into products of high quality and reliability;
(b) improved communications between the two sides during product market trials;
(c) encourage the East European partner to be more flexible in his product and delivery policies, to meet the capricious needs of the British market.

Case study no. 6 - a designer and manufacturer of automotive shock absorbers

The company which provided the information for this case study is a designer and manufacturer of shock absorbers, actively involved in licensing as part of its overseas marketing policy. The company has signed two industrial co-operation agreements for telescopic shock absorbers with Eastern European foreign trade organisations, namely a Romanian foreign trade organisation responsible for the import of industrial equipment (signed in 1966), and a Polish foreign trade organisation responsible for the import of motor industry equipment (signed in 1968). Both agreements were for a period of ten years, and both have subsequently expired. Neither of the agreements stemmed from earlier business with the Eastern European partner.

The arrangements with the Romanian partner consisted of a licensing agreement, with the provision of associated technical assistance and training of Eastern European personnel. Although the company wished to exercise quality control, access to the factory was difficult to achieve in practice. There were no associated products purchased by either side, although the company agreed to arrange for the 'buy-back' of certain Romanian-made engineering items (machine tools, electric motors, hydraulic valves) exported by another Romanian foreign trade organisation, to the value of 20 per cent of the value of the licence, (which was approximately £100,000 in total value) over a time interval of ten years. The Romanian partner had exclusive marketing rights in Romania and non-exclusive rights in other socialist countries, with the exception of Poland.

The arrangement with the Polish partner was more extensive. In addition to the granting of a licence, the provision of associated technical know-how and the training of Polish personnel (again approximately £100,000 in total value), the company sold some special purpose assembly and testing equipment also to the value of some £100,000, which is estimated to be equivalent to approximately 40 per cent of the total Polish investment for the production of these items, in British prices. The Polish partner also purchased approximately £4m of items such as seals, die-castings and bearings over the ten year life of the co-operation agreement, during which time the output of finished goods from the Polish partner increased from some £1m annually, to some £5m annually, in British prices. These items accounted for approximately 15 per cent to 20 per cent of the total value of the finished shock absorbers (in British prices); although the volume of Polish purchases has decreased as the Polish side has become more self-sufficient, or purchased some items directly from the company's suppliers. The company intended to buy Polish raw materials (e.g. zinc for die-casting) but these were not found to be competitive in price. A marketing agreement was also included, in which the Polish partner had exclusive licensing rights in Poland, and non-exclusive rights in other socialist countries. Supervision of quality control by the company was difficult to achieve in practice.

In both co-operation agreements, the licence and associated know-how were paid for on a lump sum basis. In the case of the Polish partner, this sum was paid in a relatively small number of payments, but in the case of the Romanian contract, a more complex method was followed, namely:-

30 per cent on submission of engineering documentation,
14 per cent on submission of production and testing documentation,
11 per cent on commencement of production, or 18 months after delivery of production documentation whichever was the earlier,
30 per cent in three annual instalments after the commencement of production,
15 per cent in five annual instalments up to the expiry of the agreement.

The equipment and components exported to Poland were delivered against agreed schedules, and paid for by letter of credit.

At the time of the purchase of the licences, the domestic production of automobiles using technical assistance from large Western European automobile companies, was rapidly expanding in the countries of both of the East European partners (22). It was consequently important for them to commence domestic production of shock absorbers in order to preclude the expenditure of hard currency on imports. Furthermore, the Eastern European partners also wished to manufacture a product of sufficiently advanced specification to be accepted by other car makers in international markets, the USSR in particular, and it consequently wished to obtain development cost savings due to the British company's advantage in technological expertise. In the case of the Polish partner it was also possible to draw upon the British company's availability of production capacity, and to also gain advantages from bulk buying of the company's purchased components.

The company entered into the co-operation agreements with both partners as part of its general marketing philosophy of licensing in a competitive environment, and the possibility of future licensing business with more advanced designs as the previous agreements expired (although this has not materialised in practice). In the case of the agreement with the Polish partner, it also enabled the company to secure an income from the sale of components.

Case study no. 7 - a manufacturer of computer equipment

The company described in this case study is a producer and developer of data entry and data collection equipment for use in distributed data processing systems. The company originally purchased a licence for the production and marketing of this equipment from its American designers in 1972; and followed this purchase by a considerable amount of product and market development. The areas in which the company has exclusive marketing rights for this equipment include the UK and Eire, those areas of the Far East in which sterling is the major currency used in international trade, and the socialist countries of Eastern Europe. Prior to 1972 the company was mainly engaged in the development of flight simulators and large key-to-disc applications, but considered that the market potential for these products was limited. The licence for data entry and collection equipment was consequently purchased to enable the company to enter a wider market place with a proven product.

At approximately the same time, the company decided to develop the Eastern European market, and hired a consultant for that purpose. This consultant subsequently had discussions with a number of Polish organisations, which revealed that certain of these wished to engage in joint development of a series of computer-related products, including data entry and data collection equipment. The company was also interested in some form of joint development with a Polish organisation, since it was considered to provide a means of simultaneously carrying out product and market development with a lower cost outlay. Following a series of discussions, it was finally decided to proceed with a licensing arrangement for the already proven products in the company's range. A contract was initialled to this effect in 1973, and finally signed in 1974, with the foreign trade organisation responsible for the import and export of automation equipment and measuring instruments. The licence was to be implemented in a factory which formed part of the electronic measuring division of the industrial association responsible for the production of automation equipment and measuring instruments.

The licensing agreement commenced with an initial transfer of technology, followed by three distinct stages of production development, as described below. The documentation which covered the first stage of technology transfer was originally despatched in English, with Polish translations subsequently forwarded according to an agreed schedule. These included start-ups, layouts, schematics, photography, and lists of resources and personnel required. These were despatched within three months, with subsequent appendices added later as required. In addition the contract specified training commitments by the company in terms of the number of weeks of training for defined skills (e.g. systems analysts, programmers, etc.). In addition, there was a call-off agreement on hardware, and progress payments were made through letters of credit opened at agreed phases, for amounts related to the hardware delivered and the amount of training given. The contract was progressed and monitored through project teams established on both sides, which was found to provide a satisfactory co-ordinating framework.

The first phase of the agreement was commenced in 1974, within some three months of the contract being signed, and this phase lasted for some twelve months. The company produced some twenty complete systems, and carried out the requisite assembly and testing. These systems were then dismantled into consignments of complete terminals, cables, tapes, discs, disc drives, and mini computers; and these consignments shipped separately, but carefully labelled, for subsequent re-assembly in Poland. In parallel with these shipments, the Polish partner was provided with a training system, and software development systems.

The second phase of the agreement, which ran until the end of 1976, included the delivery of kits of tape decks, disc drives and minicomputers, which had been tested independently. In addition, terminals were sent complete. The Polish customer would then assemble systems using the numbering systems provided, and consequently carry out configuration testing and system proving.

The third phase of the agreement, which ran until the end of 1978, included domestic production of all sheet metalwork by the Polish customer; and the provision of sample-tested circuit boards by the company for subsequent assembly into minicomputers by the Polish customer. At this stage, therefore, the customer was expected to carry out component testing, unit testing (usually some five days) and final configuration testing, which also usually lasted some five days. As a consequence of lack of testing expertise on the Polish side, however, there were delays in the final assimilation of production and consequent deliveries of proven computers.

From these descriptions, therefore, the agreement included the following activities:

> the granting of a licence, although this was not sold as a separate item;
> provision of associated technical assistance;
> purchase of associated components and materials;
> training of personnel;
> agreement for marketing in specified geographical areas.

The main motives for the company's interest in the co-operation agreement were the expansion of market opportunities through exploitation of the licence, and the use of the licence to provide further business contact with the Eastern European partner. The major opportunities presented were clearly the sale of technological know-how, and relevant components and units as this know-how became absorbed. In the company's view, the major motive for the Polish side's participation in the agreement was the rapid assimilation of technological know-how to meet an expanding market need for distributed data processing systems; especially as the company's systems were particularly relevant to the requirements of some of the larger Polish organisations.

A subsequent stage of the agreement was the development of a NOVA - compatible minicomputer which would accept standard interfaces used in commerce. The information transferred related to the power supply, component listings, negatives, and component specifications. In addition, three minicomputers were built, and delivered to the Polish partner to provide a physical model on which to base production. This new model was a substantial improvement over the previously licensed model in terms of memory size (64K instead of 32K) and the number of peripherals that could be operated. It was found, however, that the Polish partner had difficulties in producing a proven minicomputer of the same memory and processing speed, and similar overall compactness of design. These problems were generally caused by the Polish partner appearing to have limited access to foreign currency. Consequently, the Polish customer tended to purchase components manufactured in the socialist countries of Eastern Europe where possible, and these components were generally quite bulky. When purchasing necessary components from Western sources, the Polish customer also tended to choose the cheapest instead of those proven suppliers recommended by the company. In consequence, the processing speeds of the Polish produced minicomputers were reduced by some 20 per cent compared with specification.

The Polish partner commenced the production of these minicomputers in 1978 and the final one was delivered by the company in 1979, although the Polish partner continued to purchase basic control groups. The company also continued to provide support until the first quarter of 1980, by the shipment of terminals and power supplies purchased for important computer installations where foreign currency was available.

Case study no. 8 - a nuclear engineering company

The company described in this case study is the nuclear engineering division of a large multi-product engineering corporation. The division's main business operations are devoted to the design and manufacture of systems and equipment for civil nuclear reactors.

The company has been engaged in exploratory work and joint conferences for some eight years with almost all of the socialist countries of Eastern Europe, particularly Hungary, Romania and USSR. These discussions have centred around the general interest of all of these countries in civil applications for nuclear power, and their particular responsibilities related to the fulfillment of the integrated COMECON nuclear development programme, particularly the Soviet-led pressurised water reactor (PWR) programme. Particular interest has been shown by the Romanian side, especially in a radio-isotope plant (jointly with the company and the United Kingdom Atomic Energy Authority (UKAEA)), containment vessels and refuelling systems for PWRs.

To date, however, none of these proposals have led to contracts, with the significant exception of the sale of a radio-isotope plant to Romania, using the company's and UKAEA technology. This contract, in its turn, led to the development of a joint venture between the Romanian side and the company, with which we are more concerned in this present case study.

The contract for the radio-isotope plant was negotiated during 1974 and 1975, and included the use of a range of technologies developed by the UKAEA and the company. The construction and civil engineering aspects of the project were carried out by a Romanian contracting organisation, and the mechanical and electrical engineering was carried out by the company. The contract took some two years to negotiate in the face of strong competition from French and West German companies and its value to the company was just over one million pounds.

Following the successful completion of this project, the company submitted proposals for the project management and supply of equipment for a plant to carry out post-irradiation examination of processed nuclear fuel in Romania. The company was also facing French and West German competition for this project, and invested a high level of resources at the detailed technical negotiation stage. At the same time, however, the company was engaged in discussions with the Romanian side on two other topics, namely:

(a) a scientific and technical co-operation agreement signed at corporate level between the company's parent holding organisation and the Romanian side, relating to a range of technologies;
(b) the possibility of forming a joint venture directed by two executives from the company, and two from the Romanian Institute of Physics.

93

The objectives of this joint venture in the nuclear field were to be two-fold, namely:

(a) a primary aim of joint development in research and production equipment and systems related to the Romanian nuclear programme;
(b) a supporting activity, when the organisation had matured, of the joint marketing of project management and equipment to third countries.

The joint company was consequently established as the company viewed it as a possible vehicle to increase its opportunity for contracts in Romania. The Romanian side, in their turn, appeared to view the joint company as a vehicle for rapidly obtaining certain aspects of the company's technological know-how, and also as a means of selling Romanian engineering products in world markets.

The objectives of the joint company do not appear to have been met in practice, however, for a variety of reasons. In the first place, the company failed to obtain the contract for the post-irradiation fuel examination plant in the face of fierce French competition; and did not gain any apparent benefit from having a joint company in that event. Secondly, although not a primary aim, the company has been unable to find a market for Romanian nuclear equipment and other engineering products; whilst the Romanian side has viewed some of the company's prices for control and instrumentation equipment to be uncompetitive. Finally, although other nuclear projects have been discussed, the company has not been overwilling to invest a large amount of engineering proposal and design effort into areas where the perceived likelihood of success is far from certain; particularly as markets in other regions, especially the UK, appeared to be fairly favourable.

The company has still maintained its interest in the joint operation for a number of reasons, however; mainly related to its interest in maintaining a presence in the Romanian market place, in view of the large Romanian commitment to a nuclear power programme. This has perhaps taken on increased significance with the joint Canadian/Romanian co-operation on the CANDU nuclear project, where the company has carried out marketing efforts in both Romania and Canada.

Case study no. 9 - a manufacturer of agricultural equipment

The company described in this case study is a designer and manufacturer of equipment for use in the agricultural industry. The company has sold a number of products to each of the socialist countries, and has also acted as contractors for the provision of know-how for two package deals delivered to buyers in the German Democratic Republic. The first package deal was a factory for the production of concrete silos, in which the manufacturing equipment suppliers acted as the main contractor; and the second was a unit for the production of ten thousand head of beef, with a civil engineering company acting as the main contractor.

This case study, however, is mainly concerned with the company's transfer of technology to Hungary, through a ten year industrial co-operation agreement signed in 1968 with the Hungarian 'Komplex' foreign trade organisation. The technology to be transferred related to continuous oil fired grain driers of the company's advanced design. The Hungarian partner purchased a licence to manufacture the equipment, and

associated technical assistance to facilitate production. The company
also offered to train some Hungarian personnel, and agreed to provide
up-to-date technology over the time of the agreement. The Hungarian
purchaser had rights to market the equipment in all of the socialist
countries of Eastern Europe and in certain other countries with the
company's specific permission. Associated components and materials
were also sold by the company during the time of the agreement.

The company's main motive for entering the agreement was to provide
further business contact with its Eastern European partner. The company
had previously sold grain dryers and other equipment to Hungary, but had
reached a stage where further sales were impossible because of currency
restrictions, although the buyer still maintained a high degree of
interest in the company's products. A licensing agreement consequently
appeared to be the only way forward to remain in the market, with the
added attractiveness that the company would also obtain income from the
sale of associated components and materials.

In the company's view, the Hungarian partner purchased the licence in
order to obtain the relevant technology in a shorter time and at a lower
cost than if Hungarian organisations had carried out independent product
research and development. Furthermore, Hungary had been designated as
a major producer of agricultural equipment in the COMECON integrated
plan for national-economic development, and it was consequently important
that Hungarian agricultural engineering factories be capable of producing
advanced technology equipment to meet plan targets and obtain income
from export sales.

The company found the Hungarian partner to practice standard international procedures for licensing in terms of length of agreement,
guarantees and cancellation clauses, although there were some differences
over methods of payment. The purchaser was willing to pay cash for the
associated components and materials, but requested the company to purchase Hungarian-produced farm machinery in exchange for the licence.
The company agreed to this condition, provided that Hungarian products
of acceptable quality, delivery and price were offered. This did not
prove to be the case, however, although the company was interested in
the purchase of other goods marketed by a different Hungarian foreign
trade organisation. These goods could not be viewed as counter-purchase
by the Hungarian side, however, and a cash settlement was finally made.

Case study no. 10 - an aero-engine manufacturer

The company described in this case study is a designer and producer of
aircraft engines for use in both civil and military applications. The
company has signed two licensing agreements with Romanian organisations,
the first signed in 1972 for an engine with military applications, and
the second in 1979 for an engine to fit the airframe of a particular
type of civil aircraft. The company has also sold completed aero-
engines for use in civil aircraft, to a Romanian foreign trade organisation. The 1972 agreement extends until 1984, and the 1979 agreement
extends until 1999.

The 1972 agreement was signed by the company with the Tehnoimport
foreign trade organisation, and the Ministry of the Machine Building
Industry. The agreement covers a licence to manufacture the engine,
engine assembly, and engine overhaul; and a similar licence has also

been sold to a Yugoslavian foreign trade organisation. The Romanian and Yugoslavian partners share production of the engine components which are then exchanged so that both sides can assemble the engine, which is to be installed into a particular model of aircraft of joint Romanian/ Yugoslavian design. The licensees also have the right to fit the engine into domestically designed airframes of other types. The intention of the licensees is to gradually move to complete self-build of the engine, but major components are still being purchased from the company, together with proprietary items, which are not covered by the licence.

The licence was purchased on an initial 'lump sum' basis with a royalty payment for each aero-engine manufactured. In addition, there was a long term understanding on the supply of parts by the British company, and also a provision for the company to purchase Romanian products. These products include in order of preference: items from the aero-engine industry; engineering goods manufactured by the Ministry of Machine Building; other items manufactured by other ministries.

The 1979 agreement was negotiated with the Tehnoimportexport foreign trade organisation, but signed with the subsequently established National Aeronautical 'Centrala' ('CNA'). This organisation is part of the Romanian National Aeronautical Industry Industrial Association ('Centrala') responsible to the Ministry of Machine Building. The licensee has purchased the right to manufacture a significant proportion of the components, and to assemble the engine. The remaining components are purchased from the company and from proprietary item suppliers. The engine is then to be fitted into the airframe of a particular type of British-designed civil aircraft which is to be manufactured under licence in Romania.

The company is supplying the licensee with the requisite documentation, training and technical assistance; and the licensee can also obtain advice from the licensor on certain aspects of quality control, although it is the responsibility of the licensee to obtain civil aviation certification. The licensor also has discretionary rights to carry out quality audits within the Romanian partner's factory.

The licence was purchased on an initial lump sum basis, with a programme of purchase of components until 1985. These are priced at an agreed base price with an escalation clause based on the 'British Business Index' published by the Department of Industry. Credit facilities for the customer have been arranged by a consortium of banks with ECGD support, providing a relatively low interest rate. As in the 1972 agreement, a quantity of offset purchases was agreed, with a bias towards aero-engine and other products of the Ministry of the Machine Building Industry.

The reasons for the Romanian side entering into the industrial co-operation agreement appear to be fairly clear-cut. Romanian government policy has been directed towards the building-up of a significant aerospace industry, and Romania has chosen to turn to Western aeronautical companies for the necessary technology to produce both airframes and power units. The greater willingness and capability of the company to offer a licence, and the willingness of the UK government to permit the transfer of the technology is thought to have given the company significant advantage over potential US and French competitors. In the case of the 1979 agreement, two Romanian airlines were already successfully operating the particular aircraft-engine combination and the satisfactory service experience, together with the willingness of the UK companies to grant licences is thought to have influenced the Romanian decision.

Finally, it would appear that since the late 1960s, successive British governments have tended to view Romanian foreign policy in a comparatively favourable light. The British government has consequently tended to support the industrial co-operation agreement as an instrument for improving Anglo-Romanian relations, as evidenced by the provision of credit support through ECGD.

The company entered into the agreements to obtain income from the sale of licences and the ensuing trade in engines, components, and other goods in a market where it is frequently difficult to sell finished aero-engines. The 1979 agreement has also provided the company with the opportunity to further develop a long term relationship with Romania.

The company considered that the Romanian partner followed conventional international procedures for the length of the agreement, guarantees and cancellation clauses; although the method of payment was somewhat unusual for a licence agreement as a result of the necessity of obtaining ECGD support for the financing arrangement, and the volume of offset trade required.

COMMENTS AND CONCLUSIONS

The information obtained from the sample of British companies described in the above case studies extends the author's previously published survey on British firms' co-operation with the USSR alone (23) to include Bulgaria, Czechoslovakia, Hungary, Poland and Romania; and also extends information gained from other publications covering British experience in this field (24,25). Furthermore, the case studies cover various branches of the mechanical engineering, electrical engineering and transport equipment industries, which appear to account for some 40 per cent to 60 per cent of the total of East-West industrial co-operation agreements; and also cover the various types of co-operation activity usually associated with these industries. This chapter has also laid a basis for further discussion of the experiences of British companies in different areas of co-operation, compared with those of their counter-parts in other Western countries. These comparative experiences are discussed further in Chapter 8 below.

The companies visited during the survey described in this chapter are considered to be a typical cross-section of British engineering companies engaged in industrial co-operation with the socialist countries of Eastern Europe, during the 1970s and the late 1960s. The majority of the executives interviewed had several years of experience in trade and industrial co-operation with Eastern Europe, and consequently re-presented invaluable information sources on the practical aspects of these marketing operations. Each of the executives co-operated by providing information on the topics raised in the questionnaire, although not necessarily in an exact fashion for reasons of commercial secrecy or because the information was not always readily available. Nevertheless, it was considered that sufficient information was forthcoming to meet the objectives of the survey, and that the structured interview method served as a useful means of providing a focus for the discussion.

In every case encountered in the present survey, as in those other cases of Anglo-Soviet industrial co-operation surveyed by the present author (26), the British company found it necessary to market its products or know-how against the following background:

(a) the East European foreign trade organisation wished to arrange for the substitution of domestic production for imports, thereby meeting an industrial planning requirement with a reduced level of outlay of foreign currency;

(b) the customer also wished to arrange for the production of an item which had the capability to be competitive in the international market, and thereby generate earnings of foreign currency.

These requirements appeared to be met in almost every case encountered, although restrictions were sometimes placed on the regions into which the East European customer was allowed to export. Furthermore, in most cases it was found that the level of technological expertise of the Eastern European partner was sufficiently high to enable the purchased technology to be quickly assimilated, although there was little evidence of any substantial improvements in the purchased technology.

The British companies, in their turn, usually secured several advantages from engaging in industrial co-operation. In the first place, they secured a market for their products or expertise, where shortages of foreign currency may have precluded imports of the relevant item by the Eastern European buyer. Although technological know-how or components were sold instead of finished products, it is likely that sufficient profitability was obtained from the transaction, since most of the companies considered industrial co-operation to be an extension of their already profitable licensing business, and many of the exported components were of the high-cost, advanced-technology type. Secondly, in the majority of cases, the companies had access to relatively cheap sources of supply of certain components of adequate quality, particularly those with a comparatively high labour content. These sources of supply were frequently found to be useful when domestic production capacity was overloaded. Thirdly, the companies usually obtained closer working relationships with their Eastern European partner; an important benefit when attempting to sell in that market. Finally, payment was received in a punctual manner from the Eastern European partner.

These gains were not without disadvantage, however: Eastern European delivery requirements were usually found to be inflexible, whilst actual deliveries were frequently found to be lengthy. In addition, close control of quality was usually difficult to achieve in practice. Finally in one case, a company found itself to be in direct competition with its previous Eastern European partner, in a third market.

It is apparent, therefore, that industrial co-operation agreements are frequently far more complex to establish than the straightforward sale of a licence or capital equipment, from the viewpoint of the necessary co-ordination of aspects of design, development, production, costing and finance. For these reasons, it is probably better to base industrial co-operation agreements with Eastern European organisations on previous market knowledge in that region. Although the above complexities clearly act as a hindrance to the further development of this type of busi-

ness arrangement, it is likely that Eastern European preferences and
competitive pressures from other Western companies will require British
firms to engage more in this type of business activity if they wish to
remain in the Eastern European marketplace.

NOTES

(1) See McMillan (1977a), for a discussion of concepts and definitions
    of East-West industrial co-operation including 'inter-firm East-
    West industrial co-operation'.
(2) See Wilczynski (1976), pp.97,98.
(3) See Richman (1976).
(4) See Wilczynski (1976), pp.97,98.
(5) 'Since the weaker partner has to rise to the level of the more ad-
    vanced partner as it is a condition of mutual success' *Rynki
    zagraniczne*; 22-24/7/71; p.1 quoted by Wilczynski in note (4) above.
(6) See Hill (1978); pp.188-212, which includes a survey of 4 British
    companies engaged in industrial co-operation with the USSR.
(7) See Levcik and Stankovsky (1979); pp.216-220.
(8) See Hovyani and Ellis (1978). Hovyani pays particular attention to
    the advantages and disadvantages of industrial co-operation to
    Eastern European enterprises, and describes methods used by a
    Hungarian enterprise to select a Western partner for industrial co-
    operation.
(9) See Levcik and Stankovsky (1979); pp.44-47.
(10) Namely ECE (1978b) which included some manufacturers from the UK;
     Radice (1977) which included a survey of the experiences of eight
     British companies, and Hill (1978) (see note (6) above).
(11) Wilczynski (1976), p.86.
(12) Dragilev (1974), p.266.
(13) See McMillan (1977b).
(14) See note (6) above.
(15) The mechanical engineering (including machine tools), transport
     equipment, and the electrical engineering and electronics industries
     accounted for 36 per cent, 14 per cent and 11 per cent respectively,
     of the sample of co-operation agreements included in the ECE study
     of 1976, (see ECE (1976b)); and 22 per cent, 10 per cent and 18 per
     cent of the sample in 1978 (see ECE 1978a). The remainder of the
     sample was accounted for by the chemical industry (18 per cent in
     1976 and 26 per cent in 1978), the metallurgical industry (7 per
     cent in 1976 and 8 per cent in 1978), and a wide range of consumer
     industries (textiles, footwear, rubber, glass and furniture) and
     construction, hotel management and tourism (approximately 15 per
     cent in both 1976 and 1978).
(16) Although there are no figures available for the annual production
     and export of the actual models of mobile crane manufactured under
     licence from the British company, the Polish production of 'truck
     mounted and air-operated cranes' increased from 492 in 1965 to 1975
     in 1970, to 2310 in 1975 (see CMEA (1978), p.85). The same source
     does not, however, give import and export figures for these types
     of crane.
(17) It is apparent from the technical and economic reasons listed in the
     text, that the Hungarian partner was more interested in the inten-
     sive development of a new extraction facility, rather than the ex-
     tensive extraction of coal in general, since the total production of
     coal in Hungary did not show any increase from 1975 to 1978, and

apparently decreased compared with 1965 and 1970 figures (see CMEA
(1979), p.87).

(18) From 1975 to 1978, the Polish production of tractors increased
fairly steadily from 57,553 per year to 59,522 per year (see CMEA
(1978), p.98), although a more rapid expansion was planned for the
late 1970s and 1980s as discussed in Chapter 5 below. Consequently,
there would have been a substantially increased demand for 'original
equipment' alternators, and also subsequently for replacements. It
is difficult, however, to present similar figures for foreign trade
in Polish tractors, since CMEA statistics provide an aggregate
figure for 'tractors and agricultural implements'. Nevertheless,
from these figures, it is apparent that Poland substantially reduced its deficit for trade in these products from 71m roubles in
1975 to 3m roubles in 1978 through increased exports and reduced
imports (see CMEA (1979), pp.412-419); and there is clear evidence
from Chapter 5 below that Poland intended to increase its export
of tractors in the late 1970s and 1980s.

  Similar arguments can be advanced for the demand for car alternators, following dramatic increases in Poland's car production and
exports (see note (22) below for production and exports in 1975 and
1978).

(19) It would appear from available published statistics, that the
Hungarian partner was more interested in meeting home demand, than
overseas markets. The Hungarian production of power transformers
increased from 3,600 units in 1965 to 218,000 in 1970, although
there was a decrease to 107,000 in 1975, followed by an increase for
1977 (164,000). What is of more significance, however, is the
Hungarian output of transformers measured in capacity (2,250 MVA in
1965; 2,831 MVA in 1970, 3,519 MVA in 1975, and 5,947 MVA in 1977).
The exports of transformers by Hungary was comparatively low, falling
from 907 units in 1965 to 114 units in 1975 to 11 in 1978, although
exports could clearly increase in the future when home demands have
been met. A further indicator of the importance of home demand to
the Hungarian partner, is provided by the fact that Hungary has been
a net importer of electrical power (329 KWh exported in 1978,
compared with 4,863 KWh imported). (See CMEA (1979), pp.107,389-397
and CMEA (1978), pp.90,343,347). Much of this power has apparently
come from the USSR, and been fed to Hungary at a voltage of 750 kV,
which was subsequently transformed to 400 kV for use in the Hungarian national grid (see *Hungarian Machinery*; 1979, No. 4,pp.2-6
(M. Farkas) and pp.14-20 (V.G. Benedek)).

  Although according to official foreign trade figures the exports
of transformers by Hungary have been comparatively low, there is
also evidence to suggest that the Hungarian foreign trade organisation for electrical equipment has exported appreciable quantities
of complete generating sets including transformers. This has been
achieved through industrial co-operation arrangements particularly
to Turkey, Greece and Finland (ECE (1979a), pp.29-67).

(20) Bulgaria has been a large net exporter of material handling equipment since 1970:

	1965	1970	1975	1978
Exports (m roubles)	70.2	138	303	522
Imports (m roubles)	25.9	8.1	25.9	36.7

Source: CMEA (1978), p.337; CMEA (1979), pp.381,385.

(21) Car and truck production in Czechoslovakia during the mid 1960s and
the 1970s has been as follows:

	1965	1970	1975	1978
Car Production (units)	77,705	142,858	175,411	175,585
Truck Production (units)	16,456	24,462	33,407	40,129

Source: CMEA (1978), p.87; CMEA (1979), pp.103,104.

(22) The production and foreign trade figures for cars, relating to Poland and Romania are shown below. From these figures, it is apparent that in both countries the licensed shock absorbers would be necessary as original equipment, particularly in overseas markets; and as replacement units as the stock of automobiles increased in the respective countries.

(a) Data for Poland (see CMEA (1978), pp.87,365,369; CMEA (1979), pp.103,412-419).

	1965	1970	1975	1978
Production (units)	24,839	64,150	164,322	325,702
Exports (units)	5,485	23,837	54,500	81,229
Imports (units)	21,095	16,492	25,204	50,659

(b) Data for Romania (see CMEA (1978), pp.87,372-374; CMEA (1979), pp.103,420-422)

	1965	1970	1975	1978
Production (units)	3,653	23,604	68,013	81,375
Exports (units)	348	5,405	21,605	11,796
Imports (units)	11,880	11,451	3,989	13,181

(23) See note (6) above.
(24) See Radice (1977) and ECE (1978b).
(25) See note (6) above.

5 Large scale industrial co-operation agreements between British companies and the socialist countries of Eastern Europe

INTRODUCTION

This chapter is an account of the experiences of three large British companies that have entered into major industrial co-operation and technology transfer arrangements with the socialist countries of Eastern Europe. These arrangements, are related to one major industry in each of two socialist countries, namely the aeronautical engineering industry in Romania, and the agricultural engineering industry in Poland. Two other cases discussed in Chapter 4 above can be usefully read in conjunction with the three cases described in this chapter, namely the co-operation between the British aero-engine manufacture and a Romanian foreign trade organisation (see Case Study No. 10, in Chapter 4 above), and the co-operation between the automotive and aerospace engineering company and a Polish foreign trade organisation (see Case Study No. 3, in Chapter 4 above).

TRADE AND INDUSTRIAL CO-OPERATION BETWEEN BRITISH AEROSPACE p.l.c. AND ROMANIA

Introduction

This case study is an account of the experiences of British Aerospace p.l.c. in the marketing of freight and passenger variants of the BAC 1-11 to Romania. The case describes the activities of the company in this market over a number of years, but more importantly demonstrates the major role played by compensation trading and industrial offsets as dominant elements in the marketing mix, when exporting to the socialist countries of Eastern Europe.

Previous market experience in the Socialist Republics

The forerunner of British Aerospace p.l.c. (BAe) was the British Aircraft Corporation (BAC) formed in 1961 from the Bristol Aeroplane Company, English Electric Aviation Division and Vickers-Armstrong (Aircraft) Ltd. Two of these companies had previously sold aircraft to socialist countries with six Vickers Viscounts having been sold to China in 1960, and four Bristol Britannias to post-revolutionary Cuba. In 1962, the company also sold three second-hand Viscounts to the 'Motoimport' foreign trade organisation for use by the Polish national airline (LOT), the purchasers insisting on dealing with the manufacturer instead of an airline. The company purchased the aircraft from a reputable airline, and added a 10 per cent surcharge for its own business operations. The value of the aircraft, with associated spares, was some £1.4 million, and the whole of this sum was covered by means of countertrade in Polish ham, payment being effected independently by short-term letter of credit.

To achieve this level of countertrade the company had to negotiate with the British Government to raise the level of import quotas for Polish ham. The company, consequently, gained some seminal experience in countertrade activity. A serious attempt was made in the mid-1960s to sell Super VC-10s to Czechoslovakia but political factors appeared to prevent the purchase of a Western type of aircraft by that country's airline.

The Romanian market 1967 to 1973

The Romanian business interest commenced in 1967 when the company's Regional Sales Executive for Eastern Europe made a routine visit to Bucharest, to learn that the Romanian foreign trade organisation, 'Tehnoimport', had recently initialled a contract for the supply of Caravelle aircraft from the French manufacturer. The British executive, being of the opinion that his company's product performed better than the French aeroplane, sought an interview with the Romanian Secretary for Foreign Trade requesting approval to enter the competition. A demonstration of the BAC 1-11, 400 series, 79 to 84 seats, was consequently arranged for August of that year, and the demonstration was conducted satisfactorily. An offer was then written and presented by the company for the export of six aircraft to Romania.

During subsequent discussions with the Director-General of Tehnoimport it became apparent that the purchase decision was being influenced by a range of factors. The first of these was a need for Romanian airlines to compete in the Western tourist and scheduled markets with a modern jet; the second was the necessity to generate sufficient foreign currency from the sale of Romanian goods in Western markets to assist in meeting payment schedules for the fleet; the third and vital factor was the Romanian desire to re-create a modern aircraft industry preferably through co-operation with a Western partner. An aircraft industry had existed in Romania prior to the First World War (1) but a major factory at Brasov, which was reputed to be larger than that of Boeing in the early 1940s, had been changed over to the production of tractors after World War II. Furthermore, the Romanian industry still had access to the services of several academics with real aeronautical engineering experience - notably, Professor H. Coanda, a previous Technical Director of the forerunner to the Bristol Aeroplane Company (2); and a number of young well-qualified engineers. Neither of these groups of engineers, however, had enjoyed practical aeronautical project experience during the previous twenty years.

The sales team consequently considered that it was necessary to make a proposal which was sufficiently original to commend itself to the unique conditions prevailing in Romania. This was particularly important at a time when the French competitors were proposing a major industrial co-operation through sub-contracting the production of the Alouette helicopter.

The first attempt at an original - and uniquely British - proposal was related to an agricultural and industrial problem in the region of the Danube delta basin, where there was a factory engaged in the extraction of cellulose from reeds grown in the region. The factory was working at under-capacity, however, because of problems encountered in harvesting sufficient quantities of reeds with tractors. BAC consequently proposed to purchase for the Romanians, from within the price of the aircraft, a hovercraft from the British Hovercraft Corporation (BHC); and to promote joint technical development between BHC and the Romanian engineering industry for the design, testing and manufacture of a flail to be attached to the hovercraft and used to harvest the reeds. In this manner, harvesting of reeds would be increased, the reeds' root growth would not be destroyed for future years, the cellulose extraction factory could operate at full capacity, and some of the additional cellulose thereby produced at the factory could be sold into third countries for much-needed foreign exchange. This proposal, although energetically pursued by BHC, was not accepted by the Romanian side, however, primarily because BAC had not directed its marketing effort towards the most relevant sector of the economy. The Romanian engineering industry, which would have been responsible for the development of the system was inclined to favour the proposal, but the factory responsible for the production of the cellulose to be exported was responsible to the Ministry for the Chemical Industry! (3) The BAC 1-11s were to be bought for the account of the Ministry of Transport, thus causing extra co-ordination difficulties on the Romanian side and further preventing the project from being taken up.

At approximately the same time, however, BAC was learning to develop the concept of a 'framework contract', which temporarily linked together several independent contracts between different principals, both buying and selling. When individually concluded, these independent contracts would run in parallel and allow the extention of the framework contract.

The first principals to such an agreement were BAC and Tehnoimport, covering the purchase of the BAC 1-11 aircraft. Replacing BHC, one of the nominees in countertrade was the Britten-Norman Company based in Bembridge on the Isle of Wight, whose contract partner would also be Tehnoimport. This company, which commenced its business activities in the conversion of aircraft for use in agricultural aviation, had recently developed the twin-engined Islander aircraft to meet the specific needs of air taxi companies and commuter airlines, and other operators engaged in carriage over relatively short distances. In view of limitations on factory space, however, it was only possible to produce one aircraft per week there, whereas market demand was buoyant at ten aircraft per month (4). In the autumn of 1967, therefore, BAC arranged for Britten-Norman to meet representatives of Tehnoimport, and a production plan was negotiated. This arrangement was included in the framework contract in February 1968, and called for the Bucharest factory to deliver 215 Islanders to the following schedule (5):

First 18 months	5 aircraft
Next 12 months	25 aircraft
Each year thereafter	40 aircraft
	40 aircraft
	40 aircraft
	40 aircraft
	25 aircraft
	215 aircraft

Britten-Norman supplied drawings free of charge, major jigs at a price of some £168,000 and the requisite kit of parts at a fixed price of £8,575 each: the completed aircraft were to be delivered back to Britten-Norman at a cost of £14,775 each (6). The Romanian labour cost for assembly was thus some £6,200 per Islander, which was particularly acceptable to Britten-Norman because the contract stipulated that the assembly price was to remain fixed as long as the invoiced cost of the kits also remained fixed. This contract matched the requirements of both sides, since Britten-Norman secured production capacity of acceptable quality at reasonable costs, whilst the renascent Romanian aircraft industry was able to gain the benefit of production experience on an aircraft that was comparatively simple to build, and to plan on a steady sterling income. The Romanians built a standard Islander which was given a temporary Romanian Certificate of Airworthiness, and cleared by a British Air Registration Board surveyor based in Bucharest. These aircraft were then adapted at the Bembridge factory to meet specific customer conditions and the requirements of a full British Transport Certificate of Airworthiness (7). The total value of this offset contract was calculably £1.333 million payable to Tehnoimport for aircraft assembly work, and BAC agreed to pay a linkage fee of 9 per cent of that value (i.e. approximately £120,000) to Britten-Norman. This term 'linkage fee' is frequently used in countertrade practice to describe a fee that is sometimes paid by a Western exporter to a Western importer, who is willing to link his own purchases from the same socialist country to the seller's exports.

Another principal nominated in the framework agreement to buy from Romanian foreign trade organisations under Tehnoimport's co-ordination, was a trading company based in Switzerland which had an interest in building up its business relationship with Romania. This trading company agreed in negotiation to increase its counter-purchase obligations from $7 million to $15 million of Romanian products, with no linkage fee payable by BAC.

BAC's last nominee as principals in the framework agreement was a British machine tool importer, as the Romanian side insisted on the export of machine tools through its 'Masinimport' foreign trade organisation, responsible to the Ministry of Machine Building Industries. It was agreed that the importer would purchase Romanian machine tools worth £50,000 per year, for a period of ten years, giving a total purchase value of half a million pounds sterling, in consideration of which Masinimport granted him an exclusive agency in the UK for Romanian machine tools. The British Government quota of £35,000 per annum of Romanian machine tools was increased to £50,000 to facilitate this arrangement. There was no linkage fee to be paid but BAC agreed to introduce and absorb the machine tool imports for the first two years with a 15 per cent price mark-up to cover servicing and normal profit (i.e. some £115,000). These machine tools were subsequently used in BAC's apprentice training schools.

The framework agreement was finally signed in February 1968 following only six months negotiation and was fulfilled by conclusion of all the parallel contracts with terms of credit on the aircraft in May 1968. BAC was to sell six BAC 1-11s and spares to Romania: the first aircraft was to be delivered in June 1968, the second in December 1968, and the remaining four during 1969. The price per aeroplane was £1.6 million, and the spares price was some half a million pounds, giving a total price of £10.1 million, the £ having been devalued against the $ towards the end of 1967. 90 per cent of this sum was to be provided on export credit over seven years with ECGD backing. On the Romanian side, this total contract price of just over £10 million would be about 80 per cent covered by the total of their exports of machine tools (£0.5 million) to the British agent, the aircraft to Britten-Norman (£1.333 million), and other products to the Swiss trading company (approximately £6.25 million). Furthermore, the Romanian side received some seven years credit for their purchases, although it was usual for only short credit to be extended to their purchasers. Finally, the BAC 1-11 fleet entered the Western routes with a high earning potential for the Romanian TAROM airline. The only off-set cost to BAC associated with counterpurchase, was some £120,000 in linkage fees to Britten-Norman, and some £115,000 in machine tool purchases from the importing agency.

Although different Western parties to the framework agreement were dealing with different Romanian foreign trade organisations, the Ministry of Foreign Trade co-ordinated these parallel agreements on the Romanian side. Since BAC had originally found the Western parties to the parallel agreements, they acted as co-ordinators on the Western side, but no liability attached to BAC for the failure or success of the parallel contracts. The framework agreement was extinguished the moment that all of the formal contracts came into force and effect.

The parallel arrangements generally worked well from the viewpoint of all parties involved, although some problems were encountered. The BAC 1-11s were delivered in accordance with the schedule and are still in operation; furthermore, Britten-Norman received deliveries of Islanders to acceptable quality and cost, which were only slightly behind schedule as a result of initial familiarisation by the Romanian factory. These slight delays, however, appear to be comparatively small problems compared with those reportedly encountered with their other suppliers in the Isle of Wight (8). On the Romanian side, the aviation industry was re-established as a consequence of the training provisions and technical support from the British manufacturers. This agreement went well until Britten-Norman became financially over-extended in 1972, by which time over forty five Islanders had been delivered and many more were in build as work-in-progress.

When Price Waterhouse and Company were called in to Britten-Norman as Receivers, the Romanians were on totally unfamiliar ground, and it was necessary for them to understand quickly the British receivership system, whereby the firm's liabilities were to be consciously distinguished from its assets. The Romanian side also appeared to suffer an ideological affront, since it seemed that an important part of the current national-economic plan was to be breached, and the implicit duty of the State to provide employment could only be fulfilled in this project at the total exposure and expense of the State. Tehnoimport's situation as a trade creditor, however, was fortified firstly by the stock it was able to deliver if the Receiver chose to continue to trade, and secondly by the very advantageous terms of contract Britten-Norman had negotiated with BAC's help. Fortunately, the Receiver's rapid recognition of the Romanian assembly line as an asset and Tehnoimport's inherent skill in negotiation brought about an alliance for the purpose of selling the on-going business to Fairey Ltd.

The machine tool importer took two years delivery of machine tools,

for which he and Masinimport received an adequate income and incurred
negligible selling costs; but the Romanian industry seemed unable to
respond to orders for spares. This agency was terminated shortly after-
wards, however, when it was found that the Romanian foreign trade organisation had inadvertently granted an agency to another British machine
tool importer (9), but the countertrade requirements were deemed to have
been met, no claim to the contrary having been received from the Romanian
side. The final Western partner in the arrangements, namely the Swiss
trader, eventually purchased some $12 million of products with great
difficulty, having no disagio (i.e. linkage fee) to deploy in the face
of competitive buying.

Before completing this section there is one other sequence of negotiation that deserves comment if only because it further heralded a
coming event that had cast its shadow before. Tehnoimport had wished to
see a 'perspective', an aiming point for the future of its industry
beyond the Islanders; and the foreign trade organisation had obtained in
a side-letter to the set of 1-11 contracts, BAC's agreement to share
production of a 40-seater twin turboprop airliner with freight capability called the Type 201, which was on BAC's drawing board in 1968. This
project would be the subject of joint development and collaborative
production, conditional however on the Board of BAC deciding to proceed
with it. In the event, BAC decided not to, leaving at least a moral
vacuum of obligation to its intending partner. BAC cast about for an
alternative behicle: Handley Page, which was going through business
difficulties at that time, declined BAC's invitations; although Hawker
Siddeley's Avro Division formally agented BAC to propose a manufacturing
licence on its successful HS748 aircraft, with Rolls Royce Dart engines.
The formulation of an agreed manufacturing plan took up many months in
the early 1970s, but the commercial proposal was declined by the
Romanian side as they judged the aircraft not to represent a sufficient
technological advance in terms of production engineering, on which to
tie up their slender but growing resources throughout the next decade.
With this decision, the '201 Substitute' project lapsed - to re-emerge
in another form seven years later.

With Romania's high economic growth rate and increasing passenger
statistics, BAC considered that it must soon be a potential market for
the larger BAC 1-11 500 series, and six of these aircraft were offered
during 1970 and 1971. In Summer 1972, the foreign trade organisation
Contransimex bought a used 1-11 Model 401 from American Airlines, and
Tehnoimport a used 402 from BAC. The company was informed, however,
that larger capacity aircraft were not in the transport plan and could
only be purchased with 100 per cent compensation - to cover not only
first cost but interest as well. Britten-Norman became a party again to
the discussions as it had developed another model, the Trislander, and
was seeking production capacity; although that company was experiencing
financial problems, as described above. The sale of further BAC 1-11s
to Romania - or anywhere else - seemed impossible for a while when Rolls
Royce Limited went into receivership in 1971, but that situation was
rapidly resolved by HM Government action.

Nonetheless, the climate of business in the early 1970s seemed designed to frustrate BAC's civil aircraft marketing plans and nowhere
more than in Romania. A succession of commercial proposals were presented all with 100 per cent countertrade offered: one involved
Clarkson's Travel, the Court Line subsidiary, who would guarantee to

generate annually increasing numbers of new Western tourists to Romanian resorts which, at a nominal £50 per capita contribution to the Romanian economy, would have produced £3.12 million over five years and £11.8 million over ten years in compensation. The linkage fee that BAC agreed to pay for this deal was 'thirty bob a nob' (i.e. thirty shillings (£1.50) per new tourist generated) to cover Clarkson's cost of producing and distributing its travel brochure. For all its market share, however, Court/Clarkson went into liquidation, and BAC's proposal fell apart.

Another proposal involved not only a contract for more Islanders from Britten-Norman but also a licence for the Romanian aircraft industry to build the highly compatible Trislander. BAC proposed to acquire the rights on Romania's behalf by paying to Britten-Norman £100,000 from the proceeds of sale of each of six BAC 1-11 Model 500 Aircraft, leaving Britten-Norman to afford Tehnoimport credit for the balance of the £1.2 million licence fee. This attractive prospect and all the components of £35 million of proposed new bilateral trade came to an abrupt conclusion when Britten-Norman went into receivership in 1972 as described above - although the framework agreement was actually then under active negotiation.

Viewing with concern a general downturn in its accessible markets, BAC was considering the termination of BAC 1-11 aircraft production altogether in early 1973.

The Romanian market 1973 to 1976

By Autumn 1973, however, BAC was surprised to find that there was a prospect after all of selling the 500 Series to Romania. Tehnoimport declared itself interested in the purchase of a fleet of five such aircraft; Britten-Norman had been bought by the Fairey Engineering Company such that Islander production continued, and the Rolls Royce crisis had been solved through the formation of a new, nationally-owned company.

The value of the possible sale by BAC and Rolls Royce to Romania was estimated to be in the region of £20 million, but quite apart from the absence of an on-going production line, problems were being encountered within BAC as a precursor to the possible change into national ownership. The owners at that time were GEC and Vickers each with a 50 per cent share; but it was almost certain that the company was to be taken into public ownership, with no guarantees being apparent to the owners that they would be recompensed for any fresh investment in working capital such as that necessary to re-launch the 1-11 line to take the Romanian order. Sensing BAC's uncertainty, the Romanian side invited proposals also from the McDonell-Douglas (MDC) and Boeing aircraft companies, both of whom responded with alacrity. MDC's successful DC-9 twinjet had been kept in production over the lean years partly by US Department of Defense orders for the C9A 'Nightingale' Casualty-Evacuation variant, and partly by commercial demand for new variants. Boeing's 737 had survived with US Air Force orders for navigational trainers and sustained commercial demand. Both had benefited from their engine manufacturer's continuous development of the JT9D, while Rolls Royce appeared to have lost some time and competitive edge with the Spey, during their years of financial embarrassment and pre-occupation with RB211.

Both American manufacturers were keen to step into any market vacuum left by BAC. Boeing had already sold 707-320s to Tehnoimport for

TAROM's long-haul operations and their contracting style was therefore known and trusted, as was BAC's. Douglas had enjoyed repeated sales of the DC-9 to Yugoslavia and Austria, and brought the regional influence of both to bear on their Romanian negotiation. Market penetration at this level for either American competitor could, they thought, spell major business in future for their wide-body equipment. It was BAC's view, however, that it was in a position of relative strength as the incumbent manufacturer, and it was decided to attempt to maintain the Romanian's interest, whilst approaching HM Department of Industry for funds to cover re-start costs, which would subsequently be capitalised by the Department when the company went into public ownership. The possibilities of a further business deal between BAC and Tehnoimport thus began to fall into place during the autumn of 1974.

The negotiation predictably focused not just on price but also on the countertrade offered; what was different this time was Tehnoimport's directive from the ministries: the old insistence on 'compensation' in sundry goods to a certain value had given way to an ideological and practical preference for 'offsets' within the same sector of the economy as that to which the sale was being made. It followed that, whereas all three contenders had to offer aircraft parts manufacture, it was the calibre of that offer that would count. MDC offered the DC10 undercarriage under sub-contract, a dauntingly sophisticated task that their North American sub-contractors were finding difficult. Boeing attempted an industrial offset transaction not unlike BAC's 1968 Islander deal, but with Short's of Belfast as partners. BAC made arrangements as before but with Fairey-Britten-Norman (FBN) and offered the Romanian aircraft industry sub-contracts on future batches of BAC 1-11 components. BAC judged the mix of offset work to be competitively important, in that Islander manufacture was already within the Romanian's scope and therefore ready money in a sense; and 1-11 components incorporated technology and techniques new to the sub-contractor, which they were keen to learn.

Tehnoimport's attitude to the mix reflected the Ministry of Foreign Trade attitude: that the offset should all be in the 'ready money' category. BAC remained firm with its proposal, demonstrated its ability to import technology and loan jigs and tools free of charge as part of the package, and won for itself a firm friend and client in the Romanian aircraft industry, eventually of huge proportions.

The basis of the business deal was to be another framework agreement, which was signed on 30th December 1974, giving BAC a statement of the Romanian intent, together with a downpayment, sufficient to initiate production of the next batch of BAC 1-11s. Definitive contracts were signed in March 1975, with Tehnoimport taking 10 years export credit on the bulk of BAC's contract price but bearing no part of the start-up costs, which were covered by an advance of £6 million from the Department of Industry to be reimbursed from levies on future sales. Reciprocally, BAC was to bear no part of the Romanian learning curve or start-up costs in manufacturing the BAC 1-11 components in offset. Ironically, the Romanian industry declined to make those components for the aircraft that Tehnoimport were buying for TAROM, apparently being unwilling to bear consequential damages for any delay in deliveries to its own national carrier.

The value to BAC of the contract for 5 aircraft with spares was some £17.5 million. Their sub-contract on Romanian industry was worth about

£1.75 million; FBN committed to purchase another 100 Islanders at a unit
price of some £19,000 reflecting the Romanian industry's ability to
fabricate and assemble from raw material and proprietary items. In
addition BAC agreed to purchase some Romanian IS28 motor-gliders for
£100,000. Thus, viable offsets of over 20 per cent were created directly within the aeronautical sector. The BAC 1-11 Model 525's (see
Figure 5.1) went into TAROM's European and North African services in
1977 according to plan.

Mindful of the past history of receiverships, however, Tehnoimport
this time took care to exact from BAC a form of guarantee behind Fairey-
Britten-Norman's commitment. For reasons quite unrelated to the Romanian
production line of Islanders, Fairey Limited did indeed go into receivership in 1976 and BAC stood exposed. The Romanian side, however, never
did call the guarantees, declaring themselves 'more interested in sustaining the work than taking the money', and the parties collaborated
with the Receiver to the effect that Fairey-Britten-Norman was acquired
by Pilatus Flugzeugwerke (part of the Oerlikon-Buehrle Group) of
Switzerland; production of the Islander was then recommenced.

After three long drawn-out years of uncertainty, BAC became nationalised as British Aerospace on 29th April 1977 (10).

The Romanian market 1976 to 1979

The first inkling of further business came in October 1976 with a telex
from Bucharest enquiring what price BAC would set on a licence to manufacture the whole 1-11 airframe or, alternatively, the wings. The idea
seemed fanciful but
- as long ago as 1968 the Romanians had signalled their interest in
  manufacturing whole aircraft (the Type 201) through collaboration;
- the Romanian industry had built over 200 Islanders and were looking
  for a step-up in technology;
- the Romanian industry was already assimilating a set of twelve 1-11
  components including such items as complex as the nose wheel bay,
  control surfaces and tailplane;
- the Romanian foreign trade organisation was deeply enmeshed in
  negotiation of licences to build the German 44-seater VFW 614 twinjet (11) and its Rolls-Royce engine.

BAC declined to quote a figure for the wings only but proposed £20
million for the whole airframe. Not until March 1977 was there any
answering echo from Bucharest; it came as an invitation from the Romanian
Ambassador in London one Tuesday to outline a proposition to him by the
following Thursday and to open negotiations in Bucharest the next Monday.
This they did.

The Corporation thus embarked on a complex negotiation involving consideration of a multiplicity of possible manufacturing programmes, each
with its own price tag, whereby every part of the aircraft structure
would be transferred across to a Romanian factory in a given order. It
was quickly decided to dispense with consideration of a 'bottom-up'
licence - whereby the licensee first assimilates detail production, then
components and sub-assemblies, then assemblies and finally join-up and
flight test; and negotiations focused on a 'top-down' arrangement, in-
verting that order of work. The implications for the British lay in the
volume and value of the parts to be supplied ex-UK to the Romanian line.

The implications for the Romanians lay in the overall cost in foreign exchange and the rate at which they could practically assimilate the production processes.

On 28 May 1977, within one month of the formation of British Aerospace (BAe), a definitive protocol was signed at ministerial level which settled upon a production programme as a basis for further discussion, incorporating the 1975-contracted components and setting down the general order of manufacturing events. This programme was still to go through many mutations, each marked by signature of a suitable protocol, not least in the variants (Series 475 high-performance or Series 500 high-capacity) which the Romanians would produce, and in the number of whole aircraft BAe insisted Tehnoimport buy as a precursor to the build under licence. BAe wanted five, and in the event agreement was reached on three, in two variants, on which Romanian fitters would train at the Company's Bournemouth (Hurn) works.

Last but not least, the protocol achieved a position for the Romanians which they needed: the tenuous VFW 614 negotiation was foundering, not only because its commercial construction appeared to be ill-conceived but also because the Germans' dearth of sales seemed to belie credibility in the fundamentals of the project. The re-born BAC 1-11, in contrast, had sold both 500 and 475 Series after the TAROM batch, and there was a lively used aircraft market which BAe had cornered to good effect.

One lesson learned by the company in Eastern Europe is that a protocol, as a basis for discussion, is not an agreement; it is what it says in Marxist/Hegelian terms: a synthesis turned thesis, to which the buyer feels at liberty to produce his own antithesis, attenuating the seller's deal still further. Thus, the licence fee was still fair game for negotiation and it was to take another twelve months before this was agreed at £13.5 million, financed.

There were still some contorted negotiations to go through: BAC's manufacturing costs on 1-11 parts were declared to the licensee throughout in June 1977 US$ terms although their origin was £ sterling, and moving up with inflation. In December 1977 there was a meeting, Deputy Chairman to Minister, when problems were encountered with the dates and rate of conversion of some of the proposed figures, since BAe was, like other British exporters, under Treasury/ECGD mandate to conduct its export business in US$. The problems encountered at this phase of negotiations were used as reasons to get back into December 1977 £ sterling all round - in the knowledge that the bilateral rules of an Eastern European negotiation must sooner or later dictate that the currency of the seller be the currency of the contract.

Every aspect of the manufacturing operation came under rigorous study, involving BAe in the intense discipline of examining and re-examining its own processes and practices: definition of data pack (control drawings and supporting documentation), certification and flight test, jigs and tools, drawing standards, manpower and floorspace requirements, training and retention, technical assistance, transportation and, importantly, the interfaces with Rolls Royce and a hundred vendors of proprietary parts. Last but never least, the countertrade agreement.

In particular, the data pack, deliverable over 18 months, was to be regarded by the financial institutions as the capital export for the purposes of export credit on the fixed licence fee. This data pack was to contain all of the documentation necessary for the build of a BAe approved 1-11, with design authority to be retained by BAe during the time of the licence. The technical metamorphoses were matched by commercial contortions: in early 1978, BAe even met the Romanian request to propose a joint-venture company like that formed with Citroen for the production and counterpurchase of the Olcit car (12). BAe was not sorry to see that idea perish from over-complexity, however, since such a business arrangement seemed to be ill-suited to the structure of the aircraft industry.

The estimation and negotiation continued into the spring of 1978, when in the presence of the President of the Romanian Socialist Republic and other dignitaries, during the course of a state visit to the UK, the major definitive contracts were signed at BAe's Filton (Bristol) establishment on 15 June 1978. £1 million downpayment was made, covered in compensation by an equivalent French order for Romanian steel to be consigned to Venezuela at a disagio of 7 per cent.

Yet, still, there remained areas unsealed: finalisation of the bank-to-bank credit arrangements were dependent on a quite precise ascertainment of the values of the £100 million supply programme, and this in turn upon the definition of consignments of myriad parts, BAe and vendor. Over all cost projections lay the incalculable aspect of UK inflation, whose proportions by this time were alarming. By the time the Ministry of Machine Building Industries and the National 'Centrala' for the Romanian Aircraft Industry was satisfied with the former, the Ministry of Finance declared itself dissatisfied with the latter.

Over the ensuing months BAe pieced together a scheme of protection against excess escalation, with institutional support, to the satisfaction of the Ministry of Finance which nonetheless accepted for its account the premium involved.

By the end of 1978, however, the Ministry of Foreign Trade had expressed doubts about the level of security of its revenues, afforded by the terms of the countertrade agreement of May 1978. In this construction, BAe had accepted certain constraints as to the sectors of the economy from which goods were to be drawn. Of the £33 million counter-trade to be created at then current values, 20 per cent was to be taken from the aircraft industry - partly as 1-11 components for spares, but mostly by the production of yet further batches of Islanders (13); 60 per cent was to be taken from other engineering industries responsible to the Ministry of Machine Building Industries; the balance of 20 per cent to be taken in other commodities.

Just as Tehnoimport (now the National Aeronautical 'Centrala' (CNA)) was unable to guarantee the performance of other Romanian selling enterprises, so was BAe not obligated to guarantee the performance of buyers whom it might procure for the purposes of counter-trade. In early 1979, negotiations at the level of the Secretary of State foundered on the rock of BAe's unwillingness to put up any cash sum in forfeit against non-performance - least of all a bank guarantee as requested by the Romanian side.

On this single point, the whole carefully-wrought relationship expressed as it was with clarity in a whole inter-locking suite of contracts, simply could not be put in motion. BAe had more than consumed the down-payment in its preparatory work whilst the Romanian industry was poised for major capital investment and in a hurry to implement its plans. Rolls Royce and the vendor industries had almost ceased to believe the programme was realisable.

After negotiations had twice been broken off, a last attempt to conclude was made in May 1979 - 31 months after receipt of the initial tentative enquiry. BAe's proposition was simply an inversion of common counter-trade practice: whereas the exporter would usually provide a sum or percentage in its price structure for disbursement as disagio or linkage fees, BAe now declared its provision as £2 million and offered to deposit that sum into an account which it would open with the Romanian Bank for Foreign Trade. The rate of deposit would be pro-rata to its receipts under the sales contracts, to serve as a fund for the promotion of Romanian exports in BAe's counter-trade, its disbursement to be adjudicated jointly with CNA.

The Ministers agreed and the final signatures giving force and effect to the whole transaction were appended after midnight on 31 May 1979.

Current operations

If the main outcome of the 1975 contract in Romania was the resuscitation of the BAC 1-11 line then it can be fairly claimed that both parties benefited more than they expected at that time. BAe has taken some £250 million worth of 1-11 business since restart and its Hurn factory remains in operation; the company is therefore fully satisfied with the results of its trading relationship with Romania (14). During 1980 to 1982, the three complete aircraft were delivered after Romanian engineers and technicians had been trained in their assembly, with other training provided on major assemblies, sub-assemblies, and critical processes. In addition, sets of fuselages, wings, major sub-assemblies and components are progressively being delivered as contracted until 1986 (see Figures 5.2 to 5.6), as the Romanian partner assimilates and puts into practice the requisite know-how. By the end of 1986, twenty-two aircraft should have been produced and the Romanian partner should be self-sufficient apart from requirements for vendor parts (15).

On the Romanian side, an intense build-up of investment is evident from the appearance of the newly-built Baneasa aircraft manufacturing plant, where the major components flown out from Britain are being assembled. Although forty such aircraft are designated for Romania's own use, CNA will place emphasis on exporting its industry's product at the earliest opportunity.

It would seem, however, that the commercial conditions have not yet been created for the utilisation of the export development fund, although BAe has initiated some major ventures in counter-trade. The company has arranged the purchases of some £12 million of Romanian goods by British companies over a two year time interval, including such products and services as machine tools, fruit and travel. It is now introducing a new purchasing strategy by concentrating on larger single individual purchases of Romanian capital goods, to be sold for use on civil engineering projects in the Third World. This strategy should now facilitate a rapid fulfillment of the company's counterpurchase requirements, amounting to some £55 million to £60 million over a seven year

Figure 5.1 Initial flight of first TAROM BAC 1-11 525 (Hurn).

114

Figure 5.2 BAC 1-11 freight door under construction (Hurn).

STAGE. I
GD28 A9
L'H HALF SHELL
ASSY.JIG

Figure 5.3 BAC 1-11 fuselage half-shells under construction (Hurn).

116

Figure 5.4 Loading of completed BAC 1-11 fuselage half shells (Hurn).

Figure 5.5 BAC 1-11 build-up (Hurn).

Figure 5.6 Loading of BAC 1-11 fuselage (Hurn).

119

programme.

General conclusions

Collaborative manufacturing relationships have become almost common-
place amongst aircraft manufacturers in the present day, but BAe's
activity in Romania has some unique qualities: firstly, in that it
remains the only jet transport ever to have been licensed in its total-
ity from one national manufacturer to another; secondly, in that it
displays a consistency of management attitudes and an agility of marketing mentality over more than a dozen years of continual arms-length
negotiations, across a deep ideological divide; and finally, that in
BAe's wake, the British engine and accessory industries have found a
major new market receptive of their products for many years to come.

The positive attitude of the company towards compensation trading and
offsets has been of major significance in its business dealings with
Romanian foreign trade organisations; particularly in the development of
framework agreements between various partners, to act as a vehicle for
the development of such trade. It is this author's view that the development of such an approach has been an important beneficial element
in the company's marketing mix when exporting to that region, enabling
it to succeed against strong French, American and German competition.
There is consequently much to be learned from this company's experience,
which is relevant to the successful marketing of engineering goods and
technology to the socialist countries of Eastern Europe.

INDUSTRIAL CO-OPERATION BETWEEN MASSEY-FERGUSON-PERKINS LTD AND POLAND(16)

Introduction

Massey-Ferguson-Perkins Ltd. is a wholly-owned subsidiary of Massey-
Ferguson Ltd. It is the operating company with which the agreement was
signed for the purpose of carrying out a large scale co-operation programme with the Polish tractor industry. The emphasis was on the introduction of proven tractors and engines, and the acquisition of modern
production techniques and management systems.

The agreement related to the modernisation of the Polish tractor and
diesel engine industries focusing on the extension of the 'Ursus' group
of tractor factories based in Warsaw, by the introduction of tractors
and engines in the 38 to 60 HP category. The production capacity of
this group was to be expanded, over a fifteen year period, from the 1974
figure of 50,000 Ursus tractors per year to a potential for 75,000
Massey-Ferguson (MF) tractors per year, and approximately a further
20,000 per year of Ursus tractors in the '60 HP plus' class. The MF
tractors were to be powered by both three and four cylinder engines,
and it was intended that 98 per cent of the parts used in these tractors
would be produced in Poland.

The main agreements covering this programme were signed between the
company and the Agromet-Motoimport foreign trade organisation during
October 1974. The value of the Polish purchases of licences, associated
technology and manufacturing equipment has been estimated at some £127
million at 1974 prices (17), and formed the largest single trade and
industrial co-operation agreement between a British company and an

Eastern European socialist country at that time (18). A loan for that sum was arranged by Barclays Bank International on behalf of a syndicate of eight British banks to the Bank Handlowy w Warszawie (19), with ECGD guarantees for £104.45 million (20). The terms of repayment were up to eight years (21), at an interest rate claimed to be almost as low as 6 per cent (22); and the deal was to be completely self-financing from the Polish foreign currency viewpoint, as Massey-Ferguson-Perkins agreed to purchase some £165 million of Polish-made products. Furthermore, representatives of both the Polish and British governments signed an intergovernmental understanding in support of the project, which was considered to closely reflect the objectives of the 1973 Anglo-Polish intergovernmental agreement for economic, industrial, scientific and technical co-operation (23).

A further $228 million of credit was made available during 1978, but this loan was never taken up, although attempts are currently being made to take it up at a later date.

Activities of the agreement

The complete co-operation programme was divided into four main sections, namely:

(a) a general agreement, acting as a framework for the remaining sections of the programme;
(b) a licensing agreement relating to the product range for which licences were to be purchased, and trade mark rights were to be granted;
(c) an industrial co-operation agreement relating to the sale of CKD tractors (i.e. 'completely knocked down' kits of tractors to be subsequently rebuilt in Poland) and components to Poland, and the subsequent sale of products to Massey-Ferguson-Perkins;
(d) an implementation agreement which specified the planning, supervisory and consultancy responsibilities of Massey-Ferguson-Perkins.

The programme, therefore, included the following activities:

Purchase of licences. These purchases related not only to the traditionally made-in components of the tractor, but also those elements of a tractor that were usually bought-in by Massey-Ferguson-Perkins, such as brakes, clutches, steering gear, valves, gaskets and some hydraulic pumps. Massey-Ferguson-Perkins consequently purchased licensing rights from the suppliers of these items on behalf of their Polish partner. Those purchased items usually account for some 20 per cent by value of a total tractor.

There were also some other licences that it was unnecessary to purchase, as these had previously been purchased separately by the Polish partner, or existing Polish technology was considered to be sufficient to produce the licensed product. These related to such items as plastic and rubber moulding processes, bearings and pistons.

The provision of associated technical assistance. This technical assistance related to the provision of complete technical documentation for material specifications, process planning and associated tooling, plant layout and factory services, quality control systems and production control systems.

It is important to note that the agreement does not include the purchase of associated manufacturing equipment. As part of the planning phase of the implementation agreement the company appended its process planning documentation to the Polish partner with written associated equipment enquiries, and drew up lists of recommended suppliers. The company then vetted the proposals submitted by those suppliers, and drew up a final list of recommended purchases. The actual despatch of enquiries, together with subsequent commercial negotiations, and equipment acceptance, has been carried out by the Polish partner through the 'Metalexport' foreign trade organisation which includes machine tool purchase as one of its major responsibilities.

In addition, the Polish partner has signed two separate implementation agreements with two other British companies. These separate implementation agreements related to the foundry and forge (see the following section of this chapter) and production of fuel injection equipment (see case study No. 3, Chapter 4 above).

All civil engineering and construction work is the responsibility of the Polish partner.

The purchase of associated components and materials. The initial phase of the tractor build programme was achieved by the company providing the Polish partner with PKD ('part-knocked down') tractors which were then assembled in Warsaw. Subsequent phases of the programme consisted of part CKD delivery by Massey-Ferguson-Perkins with some units provided by the Polish partner. By the end of 1981, the agreement had reached the stage where the Polish partner was producing some 35 per cent of the tractor components, with the remaining 65 per cent delivered by Massey-Ferguson-Perkins. The final objective is for the Polish partner to domestically produce some 98 per cent of the total tractor components.

The company also entered a counter-purchase agreement to buy some £165 million of Polish produced items; this buy-back arrangement being based on the total investment value of the project. Some of these products can be sold by the company through its distribution network, and some can be used in its own factories. Other purchases include unrelated items from other Polish factories.

The exercise of quality control by the company. The Massey-Ferguson-Perkins quality and validation procedures formed part of the technology transfer. The company also retained the right to carry out any cross-checking within the Polish enterprise, to ensure that the counter-purchased products are interchangeable with those supplies drawn from other sources.

Provision of training. The provision of training by the company has been extensive, with teams of personnel from the Polish partner being trained at the company's training centre and in sub-licensees' establishments. In addition, the company has a team of four engineers located in Poland to assist their Polish partners during the implementation phases of the agreement; and the Polish partners also have an office located in the UK and staffed by Agromet and Ursus personnel.

The exchange of scientific and technical information

An agreement for marketing in specific geographical areas. The Polish partner has the right to sell licensed tractors throughout all of the COMECON countries and Mongolia. Outside this region, however, the Polish partner is only allowed to sell to Massey-Ferguson-Perkins.

Motives for the agreement. The company entered the agreement for a number of reasons, mostly related to the expansion, protection and diversification of market opportunities through the exploitation of the licence, and the use of the licence to provide further business contact with the Polish partner.

In the first place, it was apparent in the early 1970s that Poland wished to modernise its agricultural engineering industry, and was looking to Western companies for assistance in this task, including such other companies as International Harvester, Deutz and John Deere. The company had identified Poland and other Eastern European countries as growth markets for agricultural engineering products, and consequently wished to establish a protected position in that market. The role of supplier of products, components and technology seemed to be the best means of entry into the market in view of foreign currency restrictions on the purchase of finished products. Secondly, the company could see that an acceptable volume of income could be obtained from this venture through the initial sale of the licence, and subsequent sale of CKD tractors and components. Thirdly, the company realised that it could expand its international market opportunities on a competitive basis by having access to another supply source where no direct capital investment was required. Furthermore, the competitive strength of this source became more apparent for certain of those products which were being produced in Poland in high volumes, and could consequently replace other uneconomic low-volume supply sources.

The Polish partner's reasons for entering the agreement stem from its requirements to modernise its agricultural production, to become self-sufficient in food. The partner particularly required to modernise its product range of agricultural equipment, expand the volume of tractor output from 55,000 tractors per year in a country which was still reputed to use far more horses than tractors (24), and establish its agricultural engineering industry on a modern technological footing.

The option of domestic tractor product development would probably have been too lengthy within the very urgent time constraints: although the Soviet tractor industry was clearly well developed in terms of production capacity (25) it is doubtful whether that industry would have had the most advanced product technology, experience in large scale technology transfer arrangements, or the best channels of distribution for finished products, compared with advanced Western companies. Massey-Ferguson-Perkins' competitive position rested on a well tried, proven, and reliable design of product in the medium horse-power range, together with an excellent power unit. Furthermore, the company was prepared to convert its technical documentation to metric dimensions, and had many years of overseas project experience including engine licensing to Bulgaria.

Finally, Massey-Ferguson-Perkins was prepared to purchase Polish-produced items to balance the foreign currency expenditures of its partner, and British government-backed finance was available at very competitive rates, with ECGD guarantees over a sixteen year interval.

It is apparent that both sides have respectively obtained substantial benefits from the operation of the agreement. Massey-Ferguson-Perkins and its vendors have received payment of some several million pounds scheduled according to the delivery of specific packages of technological documentation, and payments have been made to both the company and its suppliers for component deliveries. Furthermore, the British machine tool industry has also benefited from the orders placed for productive equipment by Metalexport (26).

The Polish partner from its side has assimilated advanced production technology for the manufacture of contemporary tractors, with attractively low rates of interest attached to its required purchases. Furthermore, the planned volume of counter-trade purchases was set at such a level that the project should have been virtually self-financing from the foreign currency viewpoint.

It is also apparent, however, that certain problems have occurred during the implementation of the project, chiefly related to its size and technical complexity. The project is running some five years behind schedule, but this has been at a time when the construction industries throughout all of Eastern Europe have been operating under a heavy workload. In spite of these delays, however, Polish agriculture is still being provided with an advanced tractor in a shorter time span than probably would have been possible with indigenous technical development. Excellent and constructive working relationships have been developed and maintained between both parties to the agreement.

Furthermore, Western inflation in general, and British inflation in particular, has also caused some Polish purchases to be postponed, and the project consequently delayed. As a consequence of inflationary conditions, Massey-Ferguson-Perkins agreed in 1979 to increase its counterpurchases to compensate for inflation.

LICENSING AND TECHNOLOGY TRANSFER BETWEEN GKN CONTRACTORS LTD. AND POLAND

Introduction

GKN Contractors Ltd. a member of the GKN group of companies, is a large project management and engineering organisation, which has been involved in two technology transfer arrangements related to the Ursus tractor project. Contracts were signed in 1974 for the Polish purchase of forging and casting technology, and a further contract was signed in 1976 for the purchase of bimetallic bearing technology; but this section of the chapter will be concerned with the first contract only.

Prior to the signing of these agreements, the company had also signed a contract in 1973 for the supply of equipment and technology for a forge located at Jawor in the south of Poland, engaged mainly in the manufacture of components for agricultural machinery and tractors. The contract specified the transfer of technical know-how, and forging equipment including dies and tools. The requisite forging machinery capacity to be subsequently exported was estimated from a sample of seventy two representative component drawings, extended on a pro-rata basis to a total required output figure of 40,000 tons per year. In addition special tooling was provided for some fifteen specified components.

The machinery and tools were supplied F.O.B., with subsequent supervision of erection and training, and the company's main motive for entering into this business arrangement was the obvious one of income and profits from the sale of the plant. In the company's view the main motives for the Polish partner entering into the agreement were the speedy acquisition of press-forming technology and associated production capacity, which could also supplement an existing Polish competence in hammer-forging technology. The plant was sold at a fixed price with progress payments as the plant was delivered, with some countertrade purchases being made by the company's office in Poland as part of a 'best endeavours' countertrade agreement.

The Ursus forge and foundry

The business arrangement for the Ursus project consisted of a single 'software' contract of some £9 million to supply management and know-how services over a 51 month time interval for three plants, namely:
  Ursus Forge, Warsaw - 500,000 tons per year output
  Ursus Foundry II, Warsaw - 55,000 tons per year output
  Lublin Foundry - 100,000 tons per year originally, but subsequently
    amended to two phases each of 50,000 tons per year.
The projects at Warsaw consisted of the expansion and modernisation of existing facilities; whereas the project at Lublin consisted of the factory design from a greenfield site.

The company considered that it obtained this contract for two main reasons. In the first place, it was well-known to the purchasing Polish foreign trade organisation 'Metalexport' with whom it had previously signed a contract for the supply of equipment to the Jawor forge. Secondly it was a major supplier of forged parts to Massey-Ferguson in the UK, for the same range of tractors to be produced under licence in the Ursus factory; and was consequently recommended by that company in their role as major technology supplier for the Ursus group of factories.

The first stage of the agreement consisted of a study of the intended production programme, and a survey of existing production facilities to decide which equipment could be retained following refurbishment. The second stage consisted of the writing of specifications for new equipment and the listing of recommended suppliers (usually not less than three) who were then provided with enquiries. The third stage of the project consisted of the receipt of bids from suppliers, a technical and economic analysis of these bids, followed by recommendations to Metalexport. The final stage of the pre-production phase consisted of the provision of a technical support to Metalexport during subsequent discussions with equipment suppliers, although the foreign trade organisation carried out its own negotiations with suppliers and subsequently placed orders. The pre-production phase of the agreement also included the provision of recommendations for quality control and production control, and the supply of die, tool and casting drawings in sufficient detail for the tooling to be produced. Furthermore, the company also provided training to Polish personnel at the GKN group's own factories and also at local technical colleges. Quality audit has also been carried out to ensure that process specifications and quality control procedures have been assimilated.

To summarise, therefore, the technology transfer package consisted of the provision of associated technical assistance and know-how, the exercise of quality control, and the training of personnel. The company entered into the agreement as a consequence of the income and profit-ability to be obtained from the contract, in line with its business op-erations as a technical consultant and project organisation. Further-more, the Eastern European market was generally attractive at that time as a result of competitive credit facilities.

The company acted as a general contractor for the technology transfer package, with the group's forging division acting as a sub-contractor for the Ursus forge, and an American and a French organisation acting as sub-contractors for the Ursus and Lublin foundries respectively. The financial arrangements also included an escalation clause, with no specified requirements for buy-back. Finally, the company's suppliers probably received orders for approximately £100 million worth of mach-inery in 1974 prices.

In the company's view, the main motive for the Polish partner entering into the agreement was to ensure that cast and forged components of a high quality could be produced. This was considered necessary for the Polish factory to meet the quality requirements of Massey-Ferguson, and thereby sell Ursus tractors to that company's approved standards.

COMMENTS AND CONCLUSIONS

The cases described in this chapter illustrate the importance of the following two major factors in the successful large scale transfer of technology and industrial co-operation with the socialist countries of Eastern Europe:

(a) the ability of large companies to assume the status of project-leaders to co-ordinate the transfer of technology specific to their own company, and also that of their major suppliers;
(b) the ability of large companies to arrange substantial counter-purchases of items produced by the relevant socialist country, particularly those related to the technology being transferred.

In addition, the provision of government support through ECGD backed finance at competitive repayment periods and interest sales, should not be overlooked as an important factor influencing the attractiveness of the British firms' proposals to the socialist countries.

The role played by these companies also appeared to have a significant effect on the market share held by British companies for relevant products in the socialist countries concerned. For example, following the signing of the first framework contract between British Aerospace and Tehnoimport in February 1968, British aircraft exports to Romania totalled some $25 million during 1968 to 1970, accounting for all of the Romanian aircraft imports from the major Western countries in 1968 and 1969, and 74 per cent in 1970 (see Table 5.1). British aircraft exports to Romania subsequently fell to an annual figure of between $1.1 million to $1.7 million during 1971 to 1975, but following the signing of the second framework contract in March 1975 increased to $2.9 million in 1976, $32.3 million in 1977, and $6.1 million in 1978, accounting for some 12 per cent, 61 per cent and 21 per cent of Romanian imports of Western aircraft during those latter years. These British market shares in Romania during 1968 to 1970 and 1976 to 1978 were generally larger

Table 5.1

Western Aircraft Exports to Romania (1968-1978) - millions of dollars FOB
(Market Shares shown in brackets to nearest whole percent)

	1968	1969	1970	1971	1972	1973
UK	7.8(100%)	12.1(100%)	5.1(74%)	1.7(45%)	1.1(61%)	1.3(28%)
France	-	-	1.8(26%)	2.1(55%)	0.7(39%)	2.9(63%)
Italy	-	-	-	-	-	-
FRG	-	-	-	-	-	-
USA	-	-	-	-	-	0.4(9%)
Japan	-	-	-	-	-	-
Total	7.8(100%)	12.1(100%)	6.9(100%)	3.8(100%)	1.8(100%)	4.6(100%)

	1974	1975	1976	1977	1978
UK	1.4(3%)	1.5(6%)	2.9(12%)	32.3(61%)	6.1(21%)
France	5.1(9%)	17.5(67%)	16.0(66%)	19.1(36%)	20.5(71%)
Italy	-	-	-	-	-
FRG	0.3	0.3(1%)	0.1(1%)	0.1	0.1
USA	47.8(88%)	6.9(26%)	5.1(21%)	1.5(3%)	2.1(7%)
Japan	-	-	-	-	-
Total	54.6(100%)	26.2(100%)	24.1(100%)	53.0(100%)	28.8(100%)

Source: United Nations Economic Commission for Europe; *Bulletin of Statistics of World Trade in Engineering Products*, UNECE, New York, published annually.

Table 5.2

Total Western Aircraft Exports (1968–1978) – millions of dollars FOB
(Market Share shown in brackets to nearest whole percent)

	1968	1969	1970	1971	1972	1973
UK	324.4(11%)	419.7(13%)	319.7(9%)	395.7(9%)	481.4(11%)	604.8(11%)
France	203.3(7%)	251.9(8%)	307.0(9%)	268.9(6%)	340.8(8%)	360.1(7%)
Italy	71.0(2%)	115.3(3%)	86.8(2%)	122.9(3%)	126.0(3%)	137.5(3%)
FRG	98.3(3%)	84.9(3%)	115.9(3%)	85.1(2%)	106.2(3%)	199.9(4%)
USA	2312.4(75%)	2398.1(72%)	2656.4(75%)	3386.6(79%)	3015.2(72%)	4119.0(75%)
Japan	57.6(2%)	44.8(1%)	33.5(1%)	28.8(1%)	34.0(1%)	41.2(1%)
Total	3067(100%)	3314.7(100%)	3519.3(100%)	4288.0(100%)	4193.6(100%)	5462.5(100%)

	1974	1975	1976	1977	1978
UK	699.5(10%)	796.4(10%)	746.4(9%)	861.2(10%)	1902.0(15%)
France	398.3(5%)	654.7(8%)	924.6(11%)	995.9(11%)	854.7(7%)
Italy	202.5(3%)	211.9(3%)	325.7(4%)	278.8(3%)	410.0(3%)
FRG	220.3(3%)	318.3(4%)	650.1(7%)	915.1(10%)	1122.3(9%)
USA	5766.4(79%)	6171.3(75%)	6115.9(70%)	5865.8(66%)	8149.9(65%)
Japan	–	26.0(0%)	17.5(0%)	21.5(0%)	52.8(0%)
Total	7287.0(100%)	8178.0(100%)	8780.2(100%)	8938.3(100%)	12491.7(100%)

Source: United Nations Economic Commission for Europe; *Bulletin of Statistics of World Trade in Engineering Products*, UNECE, New York, published annually.

than the British market share for total world exports of Western aircraft which varied between $8\frac{1}{2}$ per cent and 15 per cent (see Table 5.2).
Although it may be claimed that British exports of aircraft to Romania
still only accounted for a very small proportion of the total world
exports of British aircraft (a maximum of 4 per cent in 1977), these co-
operation agreements were clearly extremely important in terms of trade
and technology transfer to the parties concerned.

A similar pattern is also evident for the exports of British machine
tools to Poland (see Table 5.3) following the signing of the industrial
co-operation agreement between Massey-Ferguson-Perkins Ltd., and Agromet-
Motoimport in late 1974. During 1975 to 1979, British exports of
machine tools to Poland totalled some $133.4 million, accounting for an
18 per cent market share of Polish imports of machine tools from the
major Western countries during those years, and some 9 per cent of total
world exports of British machine tools ($1,761.1 million) (27). These
exports compared extremely favourably with British machine tool exports
to Poland and the market share of Polish imports of Western machine tools,
during the 1972 to 1974 period (8 per cent) (28), and also the British
market share of total world exports of Western machine tools during 1975
to 1979 (9 per cent) (29). Before concluding this survey of Polish
imports of British machine tools, however, it is also interesting to
note that the UK held a comparatively large share (26 per cent) of
Polish imports of Western machine tools during 1969 to 1971 (30): this
may have been related to the reported Polish purchase of licences for
Leyland engines in the late 1960s (31), but further research is necessary
to clarify this point.

As a final point for discussion, it is important to note a further
large scale industrial co-operation agreement between a British company
and a foreign trade organisation from a socialist country; namely the
£115 million order for the supply of twenty two cargo ships and two
crane barges, received in November 1977 by British Shipbuilders (32).
The ships were to be delivered to the Anglo-Polish Shipping Company (APSV)
in Szczecin, a company owned jointly by the Polish Steamship Company (PZM)
and British Shipbuilders. Payment was to be made in instalments as construction took place during 1979, and the ships were then to be chart-
ered by PZM from APSV for periods ranging from thirteen and a half to
fifteen years, when the ships then became the property of PZM following
the payment of a nominal sum. British Shipbuilders were responsible for
arranging finance to enable APSV to purchase the ships, raising a Euro-
dollar loan of $65 million at nine per cent repayable by 1991; and a
second loan in US dollars made direct to APSV, with ECGD guarantee
subject to its normal limit of 70 per cent of the contract price (33).

The main objective of this contract on the British side was apparently
the securing of a large order, particularly during a period of recession
in the industry (34); and from the Polish side the securing of modern
shipping capacity with a minimum outlay of foreign exchange. The role
played by technology transfer appears to be minor, since the Polish
shipbuilding industry is not generally regarded as being technically
backward (35); and consequently, an account of this co-operation agreement has not been included in this book. It is clear, however, that
this case presents itself as an important one for further research at a
future date, in view of the small number of joint-venture companies that
have been established between British firms and organisations from the
socialist countries of Eastern Europe.

Table 5.3

Western Machine Tool Exports to Poland (1975-79)
millions of dollars FOB
(market shares shown in brackets)

	1975	1976	1977	1978	1979
UK	24.5(21%)	15.3(10%)	21.2(14%)	50.9(24%)	21.5(20%)
France	19.5(17%)	35.4(23%)	11.6(7%)	9.6(4%)	4.7(4%)
Italy	20.8(18%)	14.7(10%)	20.5(13%)	22.2(10%)	7.4(7%)
FRG	42.0(36%)	56.0(37%)	54.4(35%)	81.2(38%)	39.9(38%)
USA	7.9(7%)	13.4(9%)	8.2(5%)	24.6(11%)	1.4(1%)
Japan	1.6(1%)	17.4(11%)	40.9(26%)	26.2(12%)	30.5(29%)
Total	116.3(100%)	152.2(100%)	156.8(100%)	214.7(100%)	105.4(100%)

NOTES

(1) See Clarke, pp.2-7.
(2) Ibid.
(3) See also *The Observer* 14th January, 1979, p.24 ('Peg in a poke').
(4) See Lethbridge, D.G., 'The Islander: Problems of the Innovational
    Entrepreneur' in Hayward & Lethbridge (1975), pp.1-46.
(5) *Ibid.* p.19.
(6) *Ibid.*
(7) *Ibid.*, p.17.
(8) *Ibid.*, pp.17-19.
(9) See Rogers (1980).
(10) An account of this stage in the company's history is given in
    Gardner (1981), pp.278-288.
(11) See 'Joint Declaration of the Socialist Republic of Romania and the
    Federal Republic of Germany', Section III, 7th January, 1978, and
    'President Nicolaeu Ceaucescu and Chancellor Helmut Schmidt meet
    Romanian and Foreign Journalists' in Romanian News Agency, *Romania,
    Documents - Events*, Vol. 8, No. 2 (January 1978), pp.6,7,25.
(12) See *Le Point*, No. 494 (8 Mars 1983), pp.80,81 for a description of
    Citröen's joint venture arrangement with Romania.
(13) 315 Islanders had been produced by January. (See *The Observer*,
    14 January, 1979, p.24).
(14) *Private Communication* (Mr. Allen Greenwood, Deputy Chairman of
    British Aerospace plc).
(15) See *Flight International*, 15th March, 1980, pp.852-854 ('Romania's
    One-Elevens') for a description of the supply schedules for this
    contract.
(16) Another account of the company's co-operation agreement with Poland
    can be found in Paliwoda (1981), pp.153-167.
(17) See *Trade and Industry*, 19th September, 1974, p.591.
(18) See *Trade and Industry*, 5th September, 1974, p.478.
(19) See note (17) above.

(20) See *Moscow Narodny Bank Bulletin*, 28th August, 1974, p.2.

(21) See note (17) above.

(22) See note (20) above.

(23) See *Trade and Industry*, 19th September, 1974, p.589.

(24) The estimated stock of tractors in Poland has been estimated to have been some 300,000 in 1975 (see *Journal of Commerce*, 8th September, 1975). One estimate of the stock of horses in Poland, in 1974, given verbally to the author was some 2.2 million (i.e. some seven times higher than the stock of tractors).

(25) The Soviet tractor industry was producing almost 500,000 tractors per year by the end of 1974 (see *Narodnoe Khozyaistvo SSSR v 1975g*, p.266, in which an output figure of 499,600 is quoted for 1974). Soviet tractor exports were usually less than 10 per cent of output at that time.

(26) One source estimated expected sales to be £80 million, at a time when the total annual exports for the British machine tool industry was some £84 million, and total deliveries some £213 million (see *Moscow Narodny Bank Bulletin*, 18th September, 1974). The source quoted in note (20) above estimated that some £100 million (approximately 70 per cent of £150 million) would be delivered by British suppliers of plant and services. The final section of the present chapter provides data on the actual deliveries of British machine tools to Poland, although clearly not all of these would be for the Ursus project.

(27) Calculated from data presented in *Bulletin of Statistics on World Trade in Engineering Products*.

(28) During 1972 to 1974, British machine tool exports to Poland accounted for some $19.4 million, compared with $245.7 million of machine tool exports to Poland by UK, France, Italy, FRG, USA and Japan (based on figures from *Bulletin of Statistics on World Trade in Engineering Products*).

(29) The total world exports of machine tools by UK, France, Italy, FRG, USA and Japan was almost $20 billion during 1975 to 1979 (based on figures from the *Bulletin of Statistics on World Trade in Engineering Products*).

(30) From 1969 to 1971, British companies exported some $12.3 million of machine tools to Poland, compared with a total of some $46.9 million delivered to Poland by UK, France, Italy, FRG, USA and Japan (based on figures from the *Bulletin of Statistics on World Trade in Engineering Products*).

(31) See Gutman (1980).

(32) British Shipbuilders is the corporate body responsible for the overall management of the nationalised shipbuilding, repair and maintenance yards, having been established for this purpose under the Aircraft and Shipbuilding Industries Act of 1977. (See Report of the Comptroller and Auditor General in *Appropriation Accounts 1977-78, Vol. 2*, Class IV and Classes VI-IX, HMSO, 1979 (*House of Commons Paper 138, 1978-79*), p.vii). The value of the Polish order was reported in *Trade and Industry*, 25th November, 1977, p.378.

(33) See Report of the Comptroller and Auditor General (1979), p.ix.

(34) See Report of the Comptroller and Auditor General (1979), pp.v-x for a discussion of the steps taken by government to assist the British shipbuilding industry during a period of world recession. These included the setting up of an 'intervention fund' to reduce the losses made on orders in certain defined circumstances; and approximately half the sum earmarked for this fund was to be

allocated to yards producing ships for the Polish contract.

(35) M. Casey, a former Chief Executive of British Shipbuilders refers to a 'flourishing home shipbuilding industry' in Poland in p.3 of an unpublished paper (*The Facts about those Polish Ships*). Furthermore the Polish industry was to deliver a range of components for use in the ships, including the propellers (see *Trade and Industry*, 9th December, 1977, p.96 and *Lloyds List*, 20th October, 1979).

6 British purchase of licences from the socialist countries of Eastern Europe

INTRODUCTION

As previously explained in Chapter 2 above, the estimates that have been made of the volume of trade in licences and associated payments between the socialist countries of Eastern Europe (SCEEs) and the industrially developed Western states have suggested a predominantly eastward flow of licences and associated technological know-how, and a predominantly westward flow of payments either in hard currency, or goods. As a consequence of this, the majority of published research on East-West technology transfer has been chiefly concerned with the eastern flow of Western technology, as evidenced by the sources cited in Chapters 1, 2 and 4 above.

The research described in this chapter, therefore, aimed to extend the material provided in Chapters 3, 4 and 5 above, by providing information on the westward flow of Soviet and East European technology at the company level. This objective was achieved by carrying out studies of British companies that had purchased either product or process licences from the socialist countries of Eastern Europe. A sample of nine companies which were reported to have purchased licences from the SCEEs was obtained from the following sources:

(a) British companies cited by Kiser (1) in his published survey of the purchase of Soviet licences by Western companies;
(b) British companies cited by Wilczynski (2) in his published survey of the flow of licences in the transfer of technology between East and West;
(c) a British company cited by McMillan in his survey of Eastern European investment in the West (3);
(d) companies recommended by an official in the International Technology Group of the Department of Trade;
(e) personal contacts.

Of the nine companies originally contacted, one company reported by Kiser (4) to have purchased a Soviet licence for a device for measuring electrostatic charges did not wish to discuss any business arrangements on that topic. Another company also reported by Kiser (5) considered that its purchase of a Soviet licence for a counter device was of a very minor nature and no significant experience had arisen from it which would be useful to the current study. This, therefore, reduced the sample size to seven, including the following types of firms:

an electronics manufacturer,
a steel company,
an electrode coating company,
a process plant engineering company,
a manufacturer of rechargeable batteries,
a general iron foundry,
a foundry in an aero-engine factory.

The structured interview method was used in this sample of companies,
a copy of the questionnaire usually having been sent in advance of the
interview, to those companies which had agreed to participate in the
survey. The main topics included in the questionnaire (see Appendix C)
related to the partners in the agreement, the subject of the licence,
and the company's views for the reasons that both sides entered into the
agreement.

Subsequent interviews with executives from this sample of companies
during 1981 revealed that licence purchases had not been finally concluded in two cases, although quite lengthy discussions had been
carried out. Case studies of these companies are presented in the first
part of the next section of this chapter, nevertheless, since they
provide useful information on reasons for the inconclusiveness of such
licence negotiations. In the remaining five cases, three companies had
purchased licences from the Soviet Union, and two companies had purchased
one licence each from Bulgaria and Hungary.

It will be noted that no case is given of a British licence purchased
from either GDR, Poland, Romania or Czechoslovakia. Wilczynski (6)
cites the case of the purchase of a Polish crankshaft forging licence by
a British company; but the author found it impossible to actually trace
such a company after contacting both a leading British forge, and the
research association for the forging industry. The only evidence relating to Czechoslovakia was a decision by a British research association not to purchase a particular licence for metal rolling, following
investigation of the range of applications for the process.

THE CASE STUDIES

Case study no. 1 - an electronics company

A British company engaged in the design and manufacture of instruments
was reported by Kiser (7) to have purchased a Soviet licence for high
frequency testing equipment, but an interview with a director of the
company revealed that report to be untrue.

This company has exported several of its products to the socialist
countries of Eastern Europe for a number of years, but must approach
this market with a certain degree of circumspection since many of its
products can be used in defence applications. The company has been prepared to purchase items manufactured in the socialist countries to
promote its export trade in that region.

One such purchase was an order, in the 1960s, for some Soviet-produced
microwave attenuation measuring equipment, and another purchase was for

six automatic impedance measuring bridges, which may have had sales potential in the UK. In both cases, however, it was found that although the design concept of these instruments was sound, the practical execution was limited by the low level of Soviet component technology, thus making the products virtually unsaleable in Western markets.

Case study no. 2 - a steel company

A division of a British steel company engaged in the development of casting processes was also reported by Kiser to have purchased a Soviet licence for remelting and refining (8), but an interview with a leading process development engineer revealed that the report was not completely accurate.

The remelting and refining process is used for a variety of casting applications, forming cast material by the remelting of crude metal under a pool of molten slag. The technique can be applied in practice to the building up of large castings in sections, the joining together of smaller castings to form a larger product, and the casting of multi-ingots. It is claimed that the process can give a cleaner and sounder casting than conventional casting methods, with less likelihood of cracking at abrupt changes in section, and increased impact resistance. Furthermore, although the pre-casting costs may be higher than conventional methods because of the necessity of electrode manufacture, the overall costs are usually reduced because there is a higher quality yield in a batch of items.

The company became initially interested in the application of this method of refining in the early 1960s, but investigations revealed that the USSR had taken out a comprehensive set of patents on the process some five years earlier. Following correspondence with the Soviet foreign trade organisation responsible for the sale of licences, and discussions with Soviet specialists at technical conferences, the company's team of process development engineers was invited to visit the Soviet Union in the mid 1970s. The visit included the research institute where the process had been developed, and factories using the process.

The company was particularly interested in the applications of the process to the casting of a large engine component, since one of its manufacturing divisions had a contract for the manufacture of similar products. Following discussions with Soviet process development engineers at the research institute, a visit to a factory to view the installation used for the manufacture of those components, and discussion of process cost estimates with Soviet engineers; the company was interested in the purchase of relevant licences and certain associated manufacturing equipment.

The company consequently expressed an intention to purchase the licence, but progress on the Soviet side was slow. This slowness was probably due to problems in the co-ordination of the number of Soviet organisations involved, namely the Soviet foreign trade organisation responsible for the export of licences and associated documentation, the Soviet foreign trade organisation responsible for the sale of associated manufacturing equipment, the Soviet research institute which was the author of the licence, and the heavy engineering factory responsible for the manufacture of the equipment to be exported. By the time that the

Soviet side was ready to engage in business negotiations, some two or three years after the Moscow visit, the company had lost interest because the volume of its production of the relevant components had substantially declined, and the company had also undertaken re-organisation which resulted in less emphasis being placed on the refining process.

The visit to the Soviet Union had also led to discussion on five other groups of topics for which the process could have been applied. Discussions on the first of these areas revealed that the Soviet side was interested in licensing a process, but the British company's manufacturing division was not interested in its application. The reverse was apparent in the discussions on a second area of application with keen British interest, but a certain degree of reluctance on the Soviet side. Discussions of a third area of application were to be carried out at a later date, but these never materialised, although discussions on a fourth and a fifth application were followed up.

An arrangement was made for the company to deliver crude metal to two different destinations in the Soviet Union, within a time interval of six months. The one destination, a factory, was to convert the electrodes into component castings at a prescribed sum; and the other destination, a research institute, was to convert the electrodes into part-processed items free of charge. The company was late in its delivery of electrodes by about six months, however, and the electrodes were apparently mislaid in a Soviet port for some four months. The finished castings were finally delivered to a British port in the late 1970s, but it took a comparatively long time to obtain release from British customs because of incomplete documentation, differences in specified weights, and the fact that the company dealt with one customs area, whilst its shipping agent dealt with another. When the castings were finally received, it was found that the dimensions were not exact, although the surface and material structure appeared to be acceptable. Some other castings were still awaiting evaluation in another factory, at the time of the author's visit.

It would appear, therefore, that this licensing agreement failed to be signed as a result of a number of factors, in addition to the administrative problems encountered in the physical arrangement of the trials. On the Soviet side, a certain lack of commercial drive was apparent, evidenced by a slowness in the conduct of the technical and commercial aspects of the negotiations; although the quantity and administrative independence of the various Soviet organisations to be co-ordinated probably influenced the pace of their negotiators. On the UK side, there were various re-organisations taking place in the company as a result of changing ownership policies, a turnover of management personnel and declining demand. In addition, documents and materials were issued from several of the company's administrative and manufacturing establishments, possibly causing a certain amount of confusion to the Soviet side. Finally, the company appears to have lost interest in the application of this particular refining process, as a result of a policy decision – it employed twenty process development engineers in 1975, but the remaining team of two engineers was disbanded in 1980.

Case study no. 3 – the chemical division of a steel company

The chemical division of a large British steel company was reported by Wilczynski to have purchased a Bulgarian licence for the coating of

electrodes used in electric arc furnaces for the production of steel (9)
and an interview with the general manager of the division's electrode
coating plant revealed that this report was accurate.

Electrodes used in the electric arc process for the production of
steel are manufactured from synthetic graphite. This material has two
chief cost components, namely petroleum coke and the electrical energy
consumed during production; and both of these components have been
subject to rapid cost increases as a consequence of the increase in oil
and electrical energy prices. In 1968, graphite electrodes cost
approximately £260 per ton, increasing to about £300 per ton by 1974,
and £1,150 per ton in 1980. Electrodes vary from 20 cm. diameter x 1.5 m.
long (0.1 ton) to 60 cm. diameter x 2.4 m. long (1.2 tons).

The coating process applies a protective layer to the external surface
of the electrode to resist oxidation, and consequent consumption;
reducing electrode consumption by 17 per cent to 23 per cent for electrodes of similar diameter, and by 20 per cent to 30 per cent in other
specifically defined circumstances. The economies in use of coated
electrodes are provided by an example in the company's promotional
literature which assumes an electrode consumption of 5 kg per tonne of
hot metal produced for uncoated electrodes, reducing to 4 kg per tonne
of hot metal produced for coated electrodes; or a gross saving of 20 per
cent. Assuming that coating costs approximately 8 per cent of electrode
costs, a net saving of 12 per cent results; which for a furnace producing 200,000 tonnes of steel per year with an electrode cost of £1,300
per tonne, accounts for a saving of £156,000 per year. It is also
claimed that electrode conductivity is increased, and electrode handling
costs reduced. In order to use coated electrodes, it is usually
necessary to carry out minor furnace modification, chiefly to the
electrode clamps and sealing rings, but these modifications are usually
inexpensive.

The coating process is carried out in a series of stages; namely:

    pre-heat to prevent the formation of surface moisture;
    surface dressing to achieve a suitable surface for coating;
    spray application of aluminium;
    spray application of refractory;
    fusion.

Both spraying operations can be carried out on the same machine set-up,
as the feed and speed parameters are similar. The latter three processes
are carried out three times to achieve three coats.

The company became interested in the process in the late 1960s when it
was engaged in the business activity of supplying chemicals such as coke,
gas and other by-products, to the steel industry. It was considered
that a market existed for furnace electrode protection, but none of the
steel companies appeared to be interested in the development of a
relevant process.

In 1977, one of the company's engineers read an article in a European
steel trade journal on a Bulgarian process for electrode coating, and
arrangements were subsequently made for discussions with the Bulgarian
foreign trade organisation responsible for licence exports. Further
discussions were carried out, when the company was shown some electrodes

which had been coated, compared with others which had not; and Bulgarian engineers subsequently visited British steel companies to observe furnaces in action. The details of the licence were then agreed in 1968, when performance guarantees and a scale of fees were worked out, and the company was granted an exclusive licence for the UK.

The company then set up a pilot plant built to the licensor's specifications, and a team of some thirteen of the licensor's process specialists and technicians assisted the company in plant commissioning, product application and product evaluation. These initial proving tests did not meet the licensor's claims for the product (13 per cent to 17 per cent decrease in electrode consumption, compared with some 30 per cent), but since some of the proving tests were carried out under quite arduous operating conditions the licensee considered that there was enough evidence to demonstrate the viability of the process. Furthermore, the company had then become part of a large publicly owned corporation using some fifty arc furnaces, instead of a privately owned corporation using some twelve arc furnaces, and so its potential market had grown quite considerably.

The royalties were consequently negotiated over a period of approximately one year, and an appendix added to the licence in 1970 for renegotiated royalty payments. The licensor subsequently carried out improvements to the process to increase fusion rate such that the feed and speed parameters for spraying and fusion were similar, allowing spraying and fusion to be carried out on the same machine set-up (i.e. '1st fusion' could be carried out prior to, but at the same set-up, as '2nd spray'; likewise with '2nd fusion' and '3rd spray'). The company consequently became an exclusive licensee for this improvement in the mid 1970s. The original patent expired in the UK in 1979, and a new set of royalty payments was consequently calculated up to the time of expiry of the patent for the process improvements. The royalty payments have been based on electrode tonnage shipped from the plant, with a specified minimum royalty payment. In general, these royalty payments have usually accounted for some 6 per cent to 10 per cent of sales turnover.

Since the purchase of the original patent, the company has improved the process and product operating parameters to enable more substantial savings to be made in electrode consumption. As a consequence, the company has been able to secure for coating some 85 per cent of the total of 17,000 tons of graphite electrodes used annually in the UK. In addition, the company is currently coating some 13,000 tons of graphite electrodes per year for the Scandinavian market, which accounts for 90 per cent of electrode consumption in that region. This operation was commenced in 1974 through the purchase of an exclusive licence from the Bulgarian licensors, for exploitation of the licence know-how in Sweden, and some purchase of Bulgarian equipment. This was followed by contacts with Swedish steel makers, for trials with coated electrodes which had been exported from the UK. Since these trials were successful, the company set up a wholly-owned Swedish subsidiary, and set up a site in part of a declining steelworks with the aid of a grant from the Swedish government. The company is operated with local management and labour, and has proved to be highly successful as evidenced by the penetration of the Scandinavian market, and the company's diversification into the roll cladding market using a submerged-arc welding process.

The company has also negotiated non-exclusive licences from the

Bulgarian licensors for Canada, USA, France/Benelux, and Western Germany. Trials have been carried out in Canada, which consumes some 15,000 tons of graphite electrodes per year, and a joint venture has been set up in the centre of the Canadian steelmaking region with another operating division of the same company. This joint company intends to carry out trials for the USA market, which consumes some 150,000 tons of electrodes per year, although considerable capital may be necessary to penetrate this market because of its scattered location.

The company has also carried out successful trials in France and the Benelux countries, and is at the stage of setting up a wholly owned French subsidiary using some Bulgarian plant to penetrate the market of some 20,000 to 25,000 tons of electrodes per year. Trials are also about to commence in Western Germany, and it is intended to penetrate some 65 per cent of the German market (40,000 tons of electrodes per year, in total) within four years, although the Bulgarian licensor has already granted a licence to a German electrode manufacturer.

In conclusion, therefore, this case study can be considered to demonstrate the successful exploitation of Eastern European technology by a British company with the requisite technical expertise to assimilate the process and prove the product; and sufficient marketing and investment acumen to capture almost all of the British and Scandinavian markets.

Case study no. 4 - a plant engineering company

The organisation to be described in this case study is the iron and steel division of a British company engaged in the project management and equipment supply of plant for use in the minerals and metals industries. This company, in its turn, is part of a large international engineering and construction corporation with its headquarters in London.

The corporation was reported by Kiser to have purchased a Soviet licence for the cooling of blast furnaces (10), which proved to be accurate, following an interview with the Technical Director and a process manager of the company's iron and steel division. The licence was purchased in 1972 from the Soviet foreign trade organisation responsible for the sale of licences, through its British agents. The licence was to run for ten years, and contained some territorial limitations.

The patent to which the licence referred had been developed by a Khar'kov scientific research institute responsible for process development in the iron and steel industry. The technology covered by the licence related to blast furnace evaporative stave cooling systems, and the set of drawings and technical documentation delivered through the licensing arrangement covered two main themes, namely:

(a) the design of cast iron staves to be located inside the blast furnace, including their position and the dimensions of the cooling tubes;
(b) the design of the system which links together to form evaporative cooling around the furnace.

The company could also have purchased the staves themselves, but decided against this.

The blast furnace process is one of the oldest methods used in the iron and steel industry, to extract iron from its ore; and the industry has consequently had many years of experience in furnace cooling. The two major types of system used to achieve this are known as plate cooling, and stave cooling; and the blast furnaces in operation in Western industries tend to be divided in roughly equal numbers between both systems.

Plate cooling tended to become more favoured in certain companies in the West in the early 1970s, as a consequence of moves towards the more intensive use or 'faster driving' of the furnace, and the use of higher purity ores giving rise to 'lower slag' practices. Stave cooling, on the other hand, has remained the standard procedure in Soviet blast furnace cooling, with the use of lower purity ores and less intensive furnace driving conditions; and the Soviet industry has remained in the forefront of international stave cooling technology.

From 1972 onwards, however, Western designers of blast furnaces began to show renewed interest in stave cooling systems. They were slightly more expensive to build than their plate-cooling counterparts (approximately 30 per cent more expensive for a medium-sized £30 million to £40 million blast furnace system - the staves usually accounting for about 5 per cent of total furnace cost); but they offered more uniform furnace cooling, and the possibility of still providing a working lining to the shell of the furnace after refractory loss. Furthermore, the average blast furnace size demanded by customers began to increase, and there was little experience of plate cooling of larger furnaces within the UK industry.

Since the iron and steel division's major business operation was in the area of project management and design of advanced iron and steel production systems, it was considered necessary to have access to the most up-to-date proven technology in that area. Studies of Soviet technical literature revealed the Soviet industry to be advanced in the use of stave cooling practices, and the Soviet Union was also the only licensor for that technology. In addition, one of the division's major customers wished to purchase a furnace having a hearth diameter of 14 metres, instead of the current 11.2 metres, enabling a doubling of iron output from 5,000 tons per day to 10,000 tons per day (11) to be achieved. That customer also required stave cooling for the larger furnace, in view of the system's greater reliability in operation. These two major factors consequently made it imperative for the company to purchase the Soviet licence.

A further factor was related to the company's previous purchase of a licence from the Japanese Nippon Steel Corporation, for the design of large blast furnace systems, including the hearth, stack, and the hot-metal handling. The company intended to use the technology contained in the Japanese licence in the construction of the 14 metre diameter hearth furnace for its major customer. The Fuji Division of Nippon Steel had also used stave cooling for large blast furnaces, and were also licensees of the Soviet technology, which had subsequently been modified by them to cater for the more intensive operation and higher purity ores encountered in Japanese practice. In order to make it easier for the division to use the staves of Japanese design and manufacture, however, it was necessary to purchase the Soviet licence.

A final factor had been the long-standing trade relationships between the USSR and other divisions of the company, and other companies within the corporation. These included the sale of an ore pelletising plant to the Soviet Union in the late 1960s, and the subsequent sale of methanol plants and forging presses. The licence purchase was consequently viewed as one element of a long-term business relationship between the corporation and the USSR. The Soviet partner, in his turn, appeared to have entered the licensing arrangement to gain the obvious advantage of obtaining hard currency royalty payments from a Western market.

The company appeared to be well satisfied with its licensing arrangement with the Soviet Union. The stave cooling system is now in operation in one large furnace, and several smaller furnaces also have part systems in operation. The Soviet documentation was generally acceptable in detail and quality, although the company decided not to adopt the technology relating to evaporative cooling, since the system of forced water cooling, developed jointly by itself and a major customer, was considered to provide more sensitive control of temperature and to be more reliable for Western operating conditions. One of the company's process managers travelled to the Soviet Union for discussion with the specialists at the Soviet institute which filed the patent and to inspect staves, and also visited iron and steel plants that were using the stave cooling systems. The company also found the payment conditions to be generally acceptable which included an initial down payment when the licence was first purchased, with subsequent payments for every furnace constructed using the process. The size of those latter payments vary with the size of the furnace. In addition, the licensors organised a conference in 1975 for all furnace users and constructional engineers that had purchased the evaporative stave cooling process, in order to provide up-dated information for their licensees.

The only disadvantages encountered by the company were the impossibility of making arrangements at short notice for prospective customers to visit Soviet sites to view the licensed system in operation, and the general patent indemnity conditions, which did not appear to be very protective (i.e. the licensor only offered to pay up to 50 per cent of legal costs incurred by the licensee in defending his rights relating to the licence, up to a maximum value of 50 per cent of the royalties received). The disadvantages, however, are far outweighed by the advantages of gaining access to contemporary Soviet technology and practice in this area.

Case study no. 5 - a manufacturer of rechargeable batteries

A British company engaged in the manufacture and distribution of rechargeable batteries was reported by McMillan to have purchased a licence and engaged in a joint venture with a Hungarian foreign trade organisation (12). Subsequent discussions with executives of the company proved this report to be true, although manufacturing activities associated with the licensing agreement did not become fully operational for the reasons outlined in the subsequent paragraphs.

The company is a division of a group of companies engaged mainly in building and construction, although diversification into other product ranges also forms part of the group's policy. The company became engaged in this business activity following discussions with a business contact who was acting as an agent for Hungarian rechargeable cells, but who also wished to end this arrangement to concentrate on other products and services. The original joint venture was set up in the late 1960s, but the Hungarian partner's share was purchased by the company in 1980, when a tighter licensing agreement was signed and the company was granted sole agency rights in the UK for the product that was the subject of the agreement. The Hungarian partner is a large enterprise engaged in the development and production of medical equipment, which also has the right to engage in foreign trade. Rechargeable cells are used as power sources for certain types of medical equipment, which accounts for their development by the Hungarian enterprise.

The products which are the subject of the agreement are sealed silver-zinc rechargeable cells, purchased from the Hungarian partner in 'buttons' of five standard nominal capacities. The company then arranges these cells into appropriate configurations to meet customer requirements relating to electrical and dimensional characteristics. The processes used by the company are cell voltage equalisation, cell-welding, sleeving and packaging, and recharging. The patent covered by the licence relates to the separator used, and some of the additives used to allow sealing in a battery.

The agreement between the two sides includes a number of activities, namely the purchase of a licence, the purchase of associated components and materials, provision of associated technical assistance, the training of the company's personnel, the exchange of scientific and technical information, and an agreement for marketing in specified geographical areas - the company was allowed to sell the products throughout the world except to the COMECON region. In addition, the company could have purchased associated cell manufacturing equipment but has so far not taken up this part of the agreement, since the market volume has not been large enough to warrant investment in the automated plant necessary to achieve acceptable cost levels during cell manufacture. The present processes in the Hungarian factory are apparently less capital intensive, but lower Hungarian labour costs enable the cells to remain price competitive. In view of these problems, therefore, the company sees itself as a fabricator and distributor of rechargeable batteries to customer specification, using the rechargeable cells purchased from its Hungarian partner. The company was also engaged some years ago in the purchase of materials which were delivered free of charge for processing by the Hungarian partner. This practice has now ceased, however, in view of problems associated with the timing of materials requisitioning and recording of delivery quantities.

In addition to the rechargeable batteries, the company also markets battery chargers of its own and Hungarian designs.

The company entered into the co-operation agreement in order to diversify the market opportunities provided through exploitation of the licence. Since the company is comparatively small, it can economically meet customers' requirements of particular specification and small quantity, using relatively simple fabrication and packaging methods to build the batteries from standard cells. The company considered there was a market for this product since the silver-zinc cell is considered to have certain technical advantages compared with their nickel-cadmium counterparts. These advanges include a higher power to volume capacity, higher voltage, and lower self-discharge. In the company's view, the Hungarian partner entered the agreement in order to increase overseas sales turnover.

The company considered that the Hungarian partner followed standard international procedures in terms of payment, length of agreement, guarantees, and cancellation clauses.

Case study no. 6 - an iron foundry

The company described in this case study is a large and diverse engineering corporation which purchased a Soviet licence for a foundry process. The initial feasibility and process development studies relating to the licence were carried out by the corporation's development laboratories, and the process covered by the licence was subsequently introduced in one of the corporation's iron foundries.

The subject covered by the licence is the fluid sand process, in which moulding sand is mixed with a liquid binder, a self-hardening chemical and a foaming agent. The mixing action causes the formation of foam bubbles which separate and lubricate the sand grains, imparting fluid properties to the mixture although the liquid content of the mixture is very small. The fluidised sand may then be poured into moulding boxes containing a pattern, to form the required shape. The sand in the mould boxes can then be left to solidify since the self-hardening action commences after sand pouring. Following this solidification, in which drying is unnecessary, the pattern is then removed, and molten metal poured into the resultant cavity to obtain the requisite casting (13). This process differs from conventional casting methods, since the sand is not compacted around a pattern to achieve the requisite cavity shape, but is simply poured.

The company became originally interested in the process during 1964 when it obtained initial information on Soviet self-hardening sand technology, from the British Cast Iron Research Association (BCIRA). In 1965, the company's process development manager travelled to the Soviet Union to view the process, as a member of a delegation from the BCIRA invited by the Soviet foreign trade organisation responsible for licensing. The fluid sand process was demonstrated satisfactorily in four different locations in the USSR.

The Soviet side had been granted a patent in Belgium, and a patent application had also been lodged in the UK, although the result of this application was open to question since a description of the process had already been published. The company subsequently decided to carry out process development work of its own to test the feasibility of the process at the same time that a patent search was being carried out. The viability of the process was demonstrated after two years development work, and the soundness of the Soviet patent was also apparent, particularly with regard to the hardening agent.

The company consequently approached the Soviet foreign trade organisation to discuss licensing terms, because of their interest in using the process, and their wish to avoid the possibility of any patent infringement litigation. This latter aspect was particularly important, as the parent company did not wish to jeopardise its position with regard to the export of licences and manufacturing equipment to the USSR for use in the expanding Soviet motor industry. The company consequently entered the licensing agreement to expand its market opportunities through exploitation of the licence, and the use of the licence to provide further business contact with the Soviet partner.

The licensing agreement was signed during the summer of 1968, and included a down payment with additional royalty payments related to the tonnage produced using the process. The agreement covered the sale of the licence, associated technical assistance, and exclusive rights to the UK with the possibility of sub-licensing to other British foundries. The licence ran for ten years with the possibility of extension.

The company's specialists visited the USSR during the winter of 1968 for discussions with the Soviet author (14) of the patent, based at the Central Scientific - Research Institute for Machine-building Technology (TsNIITMash) in Moscow. All of the company's questions were answered satisfactorily, and tests were carried out in 1968 in the British company's pilot plant which had been built in the previous year. This was followed by the use of the process in a production foundry which became operational during 1970 (15). A third visit was made to the USSR during 1972 for further study of the process in three different locations.

The process has been particularly successful for the casting of large ingot moulds, bottom plates, ladles and balance weights. Sand usage per ton of metal can be reduced, and certain operating methods are easier than with rammed sand. The process has certain limitations related to the strength of the sand bond and the relatively slow hardening rate of the sand, but these do not detract from its usefulness for ingot mould castings. As a result of its successful application the company has been able to sub-licence the process to another British foundry.

The licensing practice of the Soviet partner appeared to be similar to international practice in all respects, and the Soviet partner was found to be extremely business-like, viewing the licence agreement as an important source of income. Any slowness in the arrangements were due mainly to the company wishing to test out the technology in the most comprehensive manner possible.

Case study no. 7 - a foundry in the aerospace industry

The organisation described in this case study is the investment casting plant of a British aerospace company, which purchased a Soviet process for investment casting in the late 1960s, as reported by Kiser (16). The licence related to a mechanised system of processing and transporting shell moulds for investment casting.

The investment casting process is usually carried out in the following stages:

(a) produce a steel die having a cavity of the same shape as the component to be produced, allowing for shrinkage;
(b) inject liquid wax under pressure into this cavity, and allow to set;
(c) separate the wax pattern from the die, and build up a series of similar patterns into a 'tree' joined together by wax runners;
(d) build up a 'shell' around the wax tier by a series of coatings, which are allowed to harden after each coating;
(e) melt the wax from inside the shell, and pour molten metal into the resultant cavity to solidify and form a casting;
(f) remove the shell from the outside of the casting, and dress and inspect.

The coating stage of the investment casting process consists of the application of a number of separate coats, followed by a sealing coat. The machine built under licence from the USSR contains four hundred workstations which are loaded sequentially with pre-coated wax 'trees'. The workstations are attached to a continuous chain which passes several times through two coating stations and drying units, to build up a series of further coats on the wax trees. The process is completely continuous and automatic, apart from initial loading and unloading, and is capable of progressing through a complete cycle of producing 400 shells in seven hours, or 1200 shells per day on a three shift basis.

The company's engineers originally found a reference to the process in a Soviet technical journal, and approached the Soviet foreign trade organisation responsible for licensing, for more information. Arrangements were subsequently made, during 1968 and 1969, for one of the company's engineers to visit the author of the patent (an engineer in a research institute engaged in process development for the automobile industry), and an automobile factory in which the process had been installed. It was decided to purchase the licence in view of Soviet experience in the high volume automated production of shell moulds – the Soviet installation had been operating for some ten years giving an output of one shell per minute, and it was claimed that there were seventeen other similar installations in different locations throughout the USSR at that time. Although the Soviet installations were generally casting less complex components than those to be produced by the British licensee, it was still considered that the technology was sufficiently sound in principle to be capable of transfer from the Soviet motor industry to the British aerospace industry. Furthermore, there was no proven Western design of a machine capable of automated production of such high volumes, at that time.

Following these initial investigations, the company decided to purchase the licence, and subsequently received a very comprehensive documentation package from the Soviet licensor, although some minor re-design was carried out by the manufacturers of the plant to comply with general British engineering practice. The plant was installed in 1970 and has worked satisfactorily with no visit being required from a Soviet engineer, and no further visits to the author of the licence being required from the company's engineers.

The company entered into the licensing agreement to benefit from the Soviet technological expertise in this particular process, to enable it to be provided with a high volume automated process without the associated process development costs. Furthermore, the use of the process has led to a reduction in the variation of weights of shell mouldings, compared with their hand-coated counterparts. ($\pm \frac{1}{4}$ lb compared with $\pm$ 1 lb, on an average shell of 12 lbs weight). Finally the machine has proved to be reliable and easy to service.

The machine has certain limitations in terms of the diameter and length of components that can be processed, but these limitations have not caused any great disadvantage in practice: the company only expected initially to be able to use the machine for some 50 per cent of its shell moulds, but subsequently found in practice that some 80 per cent could be produced. There have also been some problems encountered with tension and failure in the chain drive, but these were initially corrected by further experience in the use of the process; and similar subsequent problems were found to be due to overloading.

The Soviet partner appeared to enter into the agreement chiefly for the reasons of publicity and prestige in licensing the process to an advanced technology company; although hard currency was also obtained from the company's initial lump sum payment for building a machine to use the process.

The company considered that the Soviet partner followed standard procedures for the purchase of the licence. An extra payment would be required if the company decided to increase capacity by establishing another similar production line, and the company's engineers expressed the view that this would still be a worthwhile proposition if extra capacity of 1,000 moulds per day was required.

COMMENTS AND CONCLUSIONS

It is apparent from the information obtained in the above case studies that useful information has been obtained from this research to supplement the small quantity of publications in this field. Furthermore, the use of the structured interview method with industrial executives having experience of the westward flow of technology from the socialist countries through the purchase of licences, has proved to be a useful method of obtaining information on this topic, particularly at the level of the individual company; although an appreciable proportion of the information in some case studies appears to be unique to the companies concerned.

It is also apparent that the sample of companies covered by this study has been small, and has covered the purchase of licences from three socialist countries only, namely USSR, Bulgaria and Hungary; even though it is claimed (17) that Czechoslovakia, GDR and Poland have also sold an appreciable quantity of licences to the West. On the other hand, the author could find no reference to any other company which had purchased licences from the socialist countries, suggesting that this sample may be comparatively large in relation to the population of British companies with this business experience.

The sample of companies in this study may also be heavily biased towards those having business experience with the USSR, since three out of the five licence purchases studied were from the Soviet Union. Furthermore, the majority of licences in the sample were related to the processing of iron and steel, although this is probably to be expected from Kiser's figures for the Western sale of Soviet licences quoted in Chapter 2 above (i.e. 49 per cent of Soviet licences to the West have been for metallurgy). In addition, none of the licences could be considered to be at the vanguard of an advanced technology, although the licences were all fairly advanced technologically for those industries in which they were applied.

The majority of the executives interviewed within this sample found the technical information provided by their socialist licensors to be quite adequate for the technology to be transferred. Some companies, on the other hand, sometimes found the socialist licensor to be slow and bureaucratic in operation, with a lack of sensitivity towards the licensee's problems of operating in a market economy. These complaints

146

were comparatively rare, however, and should perhaps be seen in the context of the influence of bureaucracy as a fact of commercial life in the planned economies; and a background in which the licensee was sometimes slow to move, until he felt convinced of the business viability of the technology to be transferred. In spite of these problems sometimes arising from differences in political and economic structure, most companies found the business arrangements to be satisfactory.

Further research may result in finding other companies with experience in this field of business activity, but it is considered unlikely that many other company cases will be forthcoming if the view of one company executive, based on his experience in the foundry industry, is also true for other industrial sectors. It is the view of this executive that foundries in the socialist countries are now beginning to follow Western processes rather than vice versa, and that several of those processes successfully developed in the socialist countries and licensed by Western companies during the 1960s have now been replaced by other techniques. Furthermore, many of those processes were originally developed as a consequence of the specific technical and economic conditions more prevalent in the industries in the socialist countries than in their Western counterparts, thereby limiting their subsequent diffusion on an international scale.

Since the completion of the research described in this chapter, a paper has been published by Kiser (18) drawing attention to research on a similar topic reported by that author for the US Department of State in 1980 (19). Kiser's major conclusions do not appear to contradict those based on this research, however, although he does cite a further seven British engineering companies to whom either Czechoslovakian, Polish or East German foreign trade organisations claim to have sold licences, together with another eight claims for the sale of engineering licences in the UK by Hungarian foreign trade organisations. Hence, these claims should be studied further at a later date as an attempt to enlarge the present sample; although that may not result since only one of the seven cited sales to named companies was claimed to be 'current'.

NOTES

(1) See Kiser (1977).
(2) See Wilczynski (1977), pp.121-136.
(3) See McMillan (1978), pp.63-65.
(4) See Kiser (1977).
(5) *Ibid.*
(6) See Wilczynski (1977), pp.121-136.
(7) See Kiser (1977).
(8) *Ibid.*
(9) See Wilczynski (1977), pp.121-136.
(10) See Kiser (1977).
(11) See Adamson (1975) for the design specifications of both furnaces.
(12) See McMillan (1978), pp.63-65.
(13) See Brown (1970), pp.273-279.
(14) See Liass (1968).
(15) See *Foundry Trade Journal*, 29th June, 1972, pp.879-894.
(16) See Kiser (1977).
(17) See Wilczynski (1977), p.131.
(18) See Kiser (1982).
(19) See Kiser (1980), particularly pp.7-30, 56.

7 Technical co-operation agreements between British companies and the socialist countries of Eastern Europe

INTRODUCTION

This chapter surveys the experiences of a sample of British companies that have signed co-operation agreements for joint research and development, and the exchange of technical information, with industrial and scientific organisations within the socialist countries of Eastern Europe.

These arrangements are sometimes referred to as 'umbrella agreements' since they may cover a wide range of technologies, particularly when the larger multinational organisations are involved; or as 'framework agreements' since they are sometimes viewed by Western companies as a means to carry out joint discussions on a range of selected topics within a defined technical framework. This author prefers to use the term 'technical co-operation agreement', however, since the use of this term denotes that the main objective of the agreement is technologically determined. An umbrella agreement may, in practice, have trade in licences, components or machinery as its final objective (1); whilst the use of the term 'framework agreement' may be confused with the term 'framework contract', which is a business arrangement used to temporarily link together several principals to separate contracts within a defined counter-trading framework (see Chapter 5 above).

The present survey was carried out for two main reasons. The first of these was that since technical co-operation agreements were clearly an instrument for business contact between British companies and organisations within the socialist countries of Eastern Europe, they were worthy of study in any research project concerned with East-West trade and technology transfer. Secondly, detailed studies of this type of business arrangement do not appear to have been carried out by other researchers in this field, apart from two surveys of US businessmen reported by Theriot (2).

Part of Theriot's research paid particular attention to the motivations of those US companies signing technical co-operation agreements with the Soviet State Committee for Science and Technology, and concluded that that type of agreement was most frequently viewed as a means of entry into the potentially large Soviet market. One American businessman was cited who felt that the State Committee for Science and Technology was 'the appropriate vehicle for big deals' (3). This enthusiastic response has to be balanced by the view of another observer, however, who claimed that 'those businessmen who signed (technical co-operation) agreements are usually not those who sign contracts' (4). Uncertainty about the commercial value of such agreements is further illustrated by the view of another observer that only travel had resulted from the signing of his technical co-operation agreement, although travel often led to trade with the USSR (5). It is this author's view, however, that some of the pessimistic comments by those American businessmen, have to be viewed in relation to the stated objectives of technical co-operation agreements, which are not usually trade, but joint technological development and the exchange of information. Furthermore, it should be borne in mind that the main role of the State Committee of Science and Technology in the Soviet economy has not been one of foreign trade, but the co-ordination of scientific and technical development (6).

A similar range of opinions was also in evidence with regard to the transfer of scientific and technical information. One firm expressed concern that the Soviets were surreptitiously obtaining technology through visits to US plants by State Committee personnel, whilst on the other hand another observed that 'we are gaining information at a far greater rate than we are giving it' (7). Another observer, who appeared to hold a far more pragmatic view, considered the technical co-operation agreement to offer a 'preliminary opportunity to assess Soviet technical potential and marketing opportunities'. A further apparent realist observed that the main opportunity for his multinational company lay in a trade-off of US technology for research and development results from Soviet scientific-research institutes, project-technological institutes, or pilot plants (8).

THE SURVEY

The number of British companies that have signed technical co-operation agreements with organisations in the socialist countries of Eastern Europe appears to be quite small. Research by McMillan (9) strongly suggests that the Soviet State Committee for Science and Technology is more likely than its counterparts in other socialist countries to favour this kind of agreement; and consequently, in this study, attention was initially paid to announcements of British companies that had signed technical co-operation agreements with that organisation. A survey of various publications followed by discussions with British trade officials in 1977 (10) revealed that some eleven British companies had signed co-operation agreements with the Soviet State Committee for Science and Technology between 1972 and 1977. The field covered by these reported agreements included antibiotics, car tyres, aerospace technology, agro-chemicals, fibres, paints, oil drilling, computer systems, electronics, electrical engineering and precision instruments. Another paper published at approximately the same time listed seven British companies that had signed technical co-operation agreements with the Soviet State Committee, compared with seventeen companies from the Federal Republic of Germany, one from France, eight from Italy, six from Japan and fifty-three from USA (11).

Through personal contacts, the author also located two other British companies that had signed technical co-operation agreements, thereby increasing the British total to thirteen, and it is still possible that other cases exist that have not been recorded. It is this author's view, however, that the total number of British companies that have signed this type of agreement would probably not exceed twenty.

At least three of these companies signed agreements relating to chemical and polymer processes, with which we are not directly concerned in this present book, thus reducing the known population of companies to ten. Within that population of ten companies, the experiences of two had been previously described in one of the author's previous publications on export marketing of capital goods (12), although those previous accounts were chiefly concerned with the companies' other trading and technology transfer arrangements, of which the technical co-operation agreements appeared to account only for a small part. Consequently, the population of possible companies to survey was thereby reduced to eight. One of the companies within that population was an international leader in the area of measuring instrument technology; but correspondence with the company revealed its unwillingness to discuss further a technical co-operation agreement signed with the Soviet State Committee of Standards (Gosstandart SSSR), although the company claimed that the agreement was working in a satisfactory manner. Of the remaining seven companies, the following five agreed to interviews:

an electrical engineering company,
a photographic equipment company,
the National Coal Board,
an oil company,
an electronics company.

Information for the case studies in the following sections of this paper was obtained by means of structured interviews using the questionnaire shown in Appendix D below. This document was usually forwarded to the company in advance of the author's visit.

THE CASE STUDIES

Case study no. 1 - an electrical engineering company

The company described in this case study is a large electrical engineering group, which has signed an agreement for the exchange of scientific and technical information with the Soviet State Committee for Science and Technology. In addition, the company's corporate office is currently engaged in discussions with the administrative bodies responsible for technical co-operation in Czechoslovakia and Hungary. This case study, however, is concerned only with the agreement signed between the corporate board of the company and the Soviet State Committee of Science and Technology. This agreement was originally signed in 1975, supplemented in 1976 and renewed in 1981. The agreement did not stem from any earlier business with the Soviet Union, since the volume of sales to that country had been comparatively low.

The agreement has covered three main technologies, namely power engineering, communications and electronics; and the agreement relates mainly to the exchange of scientific and technical information, although joint research and development could also be included. The company signed the agreement for two major reasons, namely as a vehicle to provide further business contact with the socialist partner, and to consequently expand market opportunities. The Soviet partner, in his turn, appeared to view the agreement as a means of access to advanced technology.

Under the terms of this agreement, the company's communications division engaged in discussions with the Soviet State Committee for Television and Radio (Gostelradio), and the Ministry of Communication Equipment, on new types of colour television cameras and charge couple devices. During these discussions, particular attention was paid to Soviet requirements during the then forthcoming 1980 Olympic Games. The company was disappointed with the lack of commercial results, in spite of the very detailed technical exchanges which took place, and the apparent enthusiasm shown by the Soviet engineers for the company's equipment.

The company's electronics division did, however, obtain a contract for colour monitoring equipment to be used during the 1980 Olympic Games. This division carried out the systems design for the equipment, and also developed and manufactured the analyser; whilst the Soviet partner developed and manufactured the insertion equipment. Any problems of system compatibility were resolved for the 'Spartak' Games in 1979, which provided an ideal opportunity for a dress rehearsal for the 1980 Olympics. By that time, the equipment manufactured under this joint production arrangement was completely proven.

Exchanges of delegations of Soviet and company personnel have taken place under the terms of the agreement relating to power engineering, but there has been no movement towards a tangible objective of the exchange of information on specific technologies. Furthermore, the company's power engineering division has not received any commercial results in the form of orders.

Case study no. 2 - a photographic equipment company

The company described in this case study is a designer and manufacturer of high speed cameras and associated photographic equipment. In 1977 the company signed an agreement with the Lebedev Institute of the Academy of Sciences of the USSR, for co-operation in the field of research and development of ultra-high speed cameras of the image converter type. Such cameras, which are capable of diagnosing events in the pico-second (10^{-12} seconds) range are used mainly for laser research directed towards the production of electricity from fusion. The USSR has marked superiority in the production of image tubes, whereas the UK is stronger in the design of the required electronic driving circuits.

The agreement appeared to stem from a number of technical and business contacts between the company and its Soviet partners, over a time span of some twenty years. The company first became interested in co-operation with Soviet engineers in 1958, when a paper delivered at a conference by a Soviet scientist (13) indicated that the approaches of Soviet researchers and the company were similar with regard to electro-optical methods for high speed photography. Furthermore, it was the company's view that these technical solutions were better than those provided by American companies, at that time. Following some eight years of intensive technical development, the company began to produce advanced cameras, some of which were purchased by the USSR from 1966. This camera was considered to be technically superior to American-manufactured counterparts, and the American competitors withdrew from the international market during the late 1960s. In 1970, however, the US government representative of the Co-ordinating Committee (COCOM) of NATO exercised his power of veto over the sale of air cameras to the Soviet Union, and consequently, the company has been prevented from further sales of those products to the USSR since that date. Following the initial contacts in 1958, the company also went to a great deal of trouble to draw up an interesting programme of visits for Soviet scientists in the early 1970s, when official Anglo-Soviet relations were at a low ebb.

The company entered the 1977 co-operation agreement to generate the speedier application of new technology within the company, by attempting to join Soviet advanced technology in specific areas with the company's own know-how. Furthermore, the company hoped to extend its business activities through the sale of some of its products to the USSR and other COMECON countries.

In the view of the company, the Soviet partner entered the agreement to obtain knowledge of Western electronic design.

The agreement provides for the exchange of scientific and technical information, the co-ordination of research and development, and the training of personnel. It is also intended to engage in related trade in licences and components, since the Academy institute is interested in the purchase of some of the company's electronic drive units; whilst the company, in its turn, is interested in the purchase of Soviet-developed specialised infra-red cathode tubes.

Certain problems have been encountered in this area of the agreement, however, since these tubes are manufactured by a laboratory responsible to the Soviet State Committee of Standards (Gosstandart SSSR), and not the Lebedev Institute. A straightforward import and export transaction has not been possible since, apparently, the Lebedev Institute could not export the cathode tubes to the company through the appropriate foreign trade organisation, and thereby generate sufficient foreign currency for the purchase of the requisite quantity of electronic drive units; even though the Institute could obtain the requisite quantity of tubes from the Soviet manufacturer. The company could, on the other hand, purchase the tubes directly from a Soviet foreign trade organisation engaged in the import and export of certain types of high technology equipment and know-how; but this would not circumvent the problem of the purchase of drive units by the Lebedev Institute, since any share of the foreign currency earnings from the sale of the tubes would not be accredited to that institute, but to the Soviet manufacturer. Alternatively, the Soviet foreign trade organisation responsible for the purchase and sale of licences was willing to sell a licence for the process used to manufacture the tubes, but the company did not wish to purchase this information since its anticipated volume of output of these products was too small.

A possible way to solve these problems, which appears to be viewed with favour by the Academy Institute, is the exchange of drive units for cathode tubes; although the company is uncertain as to how a rate of exchange can be calculated between cathode tubes and electronic units, and how variations in the quantities required by either side can be accounted for. The solution of this problem is a current topic of discussion for both sides.

The company described in this case study is the Mining Research and Development Establishment (MRDE) of the National Coal Board. MRDE has signed agreements for technical co-operation with the ministries or other similar administrative bodies responsible for the coal industry in three of the socialist countries of Eastern Europe, namely Czechoslovakia (1979), Poland (1974), and the USSR (1975). In addition, the National Coal Board has a separate co-operation agreement with the Hungarian coal industry ministry.

The agreement with the Czechoslovakian coal industry relates to general exchange of information and collaboration on particular topics with a mining research establishment in Ostrava-Radvanice. The three particular topics selected for collaboration were:

(a) rock testing and behaviour in heading machines;
(b) general problems of testing, and the particular problems of testing powered supports under cyclic loads;
(c) seismic ranging, with particular reference to detection of faults ahead of the seam.

In general, the operation of this agreement appears to have been beneficial to both parties. There has been a useful exchange of information on rock-cutting with a Czechoslovakian authority, and some Czechoslovakian rock-cutting picks have been tested. In the case of powered supports the Czechoslovakian partner has more experience in the use of shield supports which is beneficial to the British side, but the MRDE has better testing facilities. The work on seismic ranging has encountered certain problems, however, as a result of differences in techniques for signal analysis used by both sides, and the existence of a series of patents held by both sides which are causing certain hindrances. Consequently, it may be more expedient to investigate certain other fields of co-operation such as coal preparation and flocculation.

The agreement is operated on an annual exchange visit basis, with the hosts covering the residential costs of the guests and the provision of a spending allowance. Visits have been arranged to pits and to equipment manufacturers on both sides, and it has proved possible to establish useful contacts with Czechoslovakian authorities. The main drawbacks that have arisen have been caused mainly through the slowness of the Czechoslovakian bureaucratic system, and the difficulty of not being able to contact individual scientists quickly by letter or telephone.

The co-operation agreement with Poland was originally signed as a consequence of Polish expertise in the building of new mines, including the design of standard modules for that purpose. The only co-operation that has actually taken place, however, has been an exchange of instruments; although there has also been a Polish proposal to supply equipment and personnel for the establishment of new pits.

The co-operation agreement with the Soviet Ministry for the Coal Industry, on the other hand, appears to have provided more tangible results. This co-operation agreement is administered by a steering committee which meets annually, and a series of working groups which meet once or twice each year, the frequency of meetings being influenced by the demands of the project. These working groups cover tunnelling machinery, reliability testing, dust control, computer applications, and software systems. The majority of co-operation has taken place in the areas of tunnelling machinery and reliability testing, and the remainder of the case study will describe those activities. In addition, it is considered that the Soviet industry has particular expertise in the use of hydraulic mining, and long wall techniques for steep-sided seams. Neither of these areas of work are of particular current interest to the British side, however.

The co-operation in tunnelling machinery has led to technical collaboration in the development and joint prototype production of a particular type of road-heading machine. At the time of drawing up the specification in 1975, it was intended that the machine would be of an advanced type, particularly with regard to its power. In order to achieve a currency-free method of technology exchange, it was decided that the machine could be divided into two major parts, namely the cutter unit which makes up the top half of the machine and the feeder unit which makes up the bottom half. The Soviet side would design and build two sets of the feeder unit and the British side would similarly design and build two sets of the cutter unit; exchange of one cutter unit with one feeder unit would then take place to provide each side with a completed machine. The cutter unit is made up of a cutter head, transmission unit, and a hydraulic powerpack which drives both the cutter and feeder units.

Design and manufacturing co-ordination has been mainly achieved through meetings of engineers from MRDE and the Moscow Central Scientific Research Institute of Underground Engineering (TSNIIPodZelMash). The location of these meetings alternates between Moscow and MRDE.

The project has been aided by the fact that both teams of engineers have attempted to make the collaboration work, and there has also been a continuity in the personnel making up both teams. The project has been hindered by the long distances between the two locations, language problems, and the difficulties of gaining immediate access to the relevant Soviet engineers. Furthermore, changes in the international political climate have sometimes caused the project to be halted.

The project has now reached the stage where the cutter units are being built in one of the Coal Board's main equipment suppliers, whilst the feeder units are also being built in a Soviet factory, and it is intended that units will be exchanged in 1982. As a result of the long time interval during which the project has been in operation, however, the machine will probably be out of date as British equipment manufacturers have developed more advanced machinery in the meantime; although the Soviet industry does not appear to have developed such a powerful machine.

The collaboration on reliability testing grew from initial discussions in the Anglo-Soviet Working Group on the Reliability of Mining Machinery, where it appeared that the State Project Institute for Mining Machinery (GIProUgol'Mash) in Moscow had developed advanced techniques for reliability measurement, whilst MRDE had better testing facilities and practical expertise for the evaluation of these techniques. It was decided, therefore, to test out these techniques on coal shearer gearheads, which are critical assemblies influencing the reliability of these machines. Furthermore, they account for between a quarter and a third of the total cost of shearer units.

The Soviet techniques consisted, in principle, of commencing a test at a 'unit-load' and then increasing the test load in a series of eight equal load increments to a level equal to double the 'unit-load'. The cycle of increasing and decreasing the load is run over a time of about six hours; the total duration of the test should be about 250 hours, depending upon the item being tested. The 'unit-load' in British terminology would be considered as the nominal working load which loads the most highly stressed component within the gearbox at 50 per cent higher than its rated fatigue level.

This accelerated testing approach would approximately halve the time taken for type testing in MRDE, which consists of running the gearbox at full nominal load for a period of 1,000 hours plus the associated setting-up and inspection time. This type testing method has been in operation for about ten years, and is accepted by equipment manufacturers and users alike; consequently any accelerated testing procedure would need to be thoroughly tested and correlated against existing procedure.

The main points at issue were the assumptions underlying the load cycle, namely that life and load intensity are linked by a fifth-power relationship. These assumptions were tested by comparing the characteristics of gearboxes loaded steadily for 24 hours, against those undergoing accelerated testing for 6 hours. The Soviet partner carried out five tests on five shearer gearheads (4 accelerated testing and one steady load) to verify repeatability, and the British side carried out six tests on three different types of gearbox. The test information was then exchanged, and found to be beneficial to both sides, particularly as it appears that testing times can be halved for these products. The next stage in collaborative development is to develop an acceptable testing method for pumps operating on contaminated fluids, to determine whether certain pumps are less prone to breakdown.

Case study no. 4 - an oil company

The company described in this case study is a large British multinational corporation engaged in the oil and chemicals business, which signed a five-year technical and scientific co-operation agreement with the Soviet State Committee for Science and Technology, in the 1970s.

This umbrella agreement with the State Committee was appended by three subsidiary agreements with, respectively, the Ministry of the Refining and Petrochemical Industries; the Ministry of the Oil Industry; and the Chief Administration of the Microbiological Industry, which manages, amongst other things, the USSR's synthetic animal feed industry. The first of these agreements, which was related to refining, has been concerned with two main areas of work, namely, the exchange of information on general operating procedures and maintenance at refineries, and trials related to certain processes. The company has found the discussions on refinery operating practice to be generally useful, and the examination of processes has helped to familiarise each side with the other's capabilities and highlighted interesting differences in feedstocks and specifications.

The second agreement related to oil field operations, particularly the general problems of recovery, especially from reserves with heavy oils. The flow of information to date from the company has included data on thermal recovery pilot studies, operating procedures in perma-frost conditions (based on the company's experience in Alaska) and advice on enhanced recovery techniques. The company has obtained some comparable Soviet data on thermal recovery methods, and seeks further information on the success of Soviet methods of recovery in certain defined environments.

The final agreement related to the manufacture of synthetic proteins using technology which the company had installed in a Western European plant. Some useful exchange of information had been obtained and it was intended that a unique technology would be sold to the Soviet partner for the construction of a synthetic protein plant similar to that operated by the company. This contract did not come to fruition, however, as the company decided to withdraw from this particular avenue of synthetic production, for policy reasons.

The agreements, therefore, include the exchange of scientific and technical information as the major activity. It is also hoped to carry out some co-ordinated research and development, particularly on production methods for enhanced recovery, and the trading of licences is also still a possibility.

The company appeared to enter into the agreement mainly to widen its technical contacts with the Soviet side, and also to expand its market opportunities, including the sale of licences. In addition, the company hoped for speedier application of new technologies within the company, particularly in the area of advanced oil recovery techniques. On balance, the company has found the discussions and business contact to be useful, and it has been agreed to extend the agreement for a further five years.

The Soviet side appeared to enter the agreement for similar reasons. In their case, however, the emphasis is rather on the application of technologies than on the expansion of market opportunities. On a more practical level, Soviet specialists appeared to value the opportunity to discuss their concepts and practices with their western counterparts, to obtain either substantiation or alternative viewpoints.

Case study no. 5 - an electronics company

The company described in this case study is a well established manufacturer of electronic components and finished products. The individual businesses of the company are, in the main, formed into groups, and the major geographical areas served by the company are organised into sales and marketing regions.

The company's business contacts with the socialist countries of Eastern Europe are the responsibility of one regional sales manager. Sales of licences have been made to Bulgaria, GDR, Poland, Romania and the USSR; and the company has also increased its presence in the socialist countries (except the GDR) through direct product sales, and third country co-operation through one foreign trade organisation. In addition, the company has been engaged in continuing discussion with the socialist countries, attempting to define precisely the areas of technology transfer and industrial co-operation which may be of mutual interest. This case study, however, deals with co-operation in one country only, namely the Soviet Union, paying particular attention to the company's agreement on technological co-operation with the Soviet State Committee for Science and Technology.

The agreement was originally signed in 1968 at approximately the same time that the UK signed its first inter-governmental agreement with the USSR for co-operation in the fields of science and technology. The main technical areas to be covered by the agreement were quite broadly defined, relating mainly to communications and electronic components. In 1974 the company decided to appoint a well-established Western firm with an office in Moscow, as its representative for the Soviet market; and this arrangement operated for some six years.

The agreement related chiefly to the exchange of scientific and technical information. It was signed by the company to provide further business contact with the socialist partner, and through the agreement, to provide expanded and diversified market opportunities. From the Soviet viewpoint, the agreement provided the opportunity to gain access to contemporary technical information.

It is difficult to assess the value of this agreement to the company's level of sales in the USSR, although the company has been provided with several opportunities to conduct seminars and other sales-related presentations in the Soviet Union through the terms of the agreement. The company has sold a separate licence to a Soviet foreign trade organisation, with associated know-how and plant for the manufacture of components; but the value of this contract was less than £1 million. It is not unreasonable, therefore, to question whether the order might also have been obtained even if the company had not previously signed the technical co-operation agreement with the State Committee. In addition, there has also been a contract for the supply of communication equipment, but this was also less than £1 million in value.

It is possible, however, that the company's trade possibilities with the USSR have been influenced by the nature of the company's product range, which includes items requiring export licences. In one instance at least, an export licence was refused by the British Government at an advanced stage of negotiations even though there had been little indication of such an action during earlier stages of the discussions. Consequently, it may not always be easy to create confidence amongst Soviet buyers in case export licences are not granted; and furthermore, the company's operating divisions may often choose to concentrate their marketing resources into those areas likely to yield a greater degree of success in a shorter time scale.

COMMENTS AND CONCLUSIONS

The information contained in the case studies described in this paper, raises several questions relating to the operation of technical co-operation agreements with the socialist countries of Eastern Europe.

In the first place, it is apparent that all of those companies which expected the signing of the co-operation agreement to lead directly to trade have been disappointed; since the subsequent volume of trade was generally quite low. In hindsight, this is not surprising, since the

activities relevant to the technical co-operation agreement have been
carried out with either a state committee for science and technology or
an industrial ministry, the majority of which do not possess the right
to engage in foreign trade activities on a large scale. It is likely
that officials within these organisations were far more interested in
obtaining information which related to long-term and medium-term technical policies and production strategies, rather than short- to medium-
term plant purchasing decisions. In addition, some companies entered
these agreements when Anglo/Soviet inter-governmental agreements were at
their most friendly during the late 1960s and mid 1970s, which raises
the question as to whether they may have been signed to also assist
governments' political objectives, rather than just industrial activity
between the company and its business partner.

It is this author's view, however, that the marketing effort put into
these agreements by the respective British companies should not necessarily be regarded as merely wasted. In the first place, the companies
built up channels of communication and points of contact with several
officials in industrial ministries and foreign trade organisations in
the socialist countries, which are important benefits when trying to sell
to that market, since the end-user is frequently organisationally separated from the purchaser. Consequently, in many ways, the operation of
a technical co-operation agreement can be viewed as a medium to long
term promotion exercise.

It is also important to note, that some companies considered the
signing and operation of the technical co-operation agreements to be
extremely beneficial. This was particularly the case where it was apparent that the objectives of the agreement were clearly technical and
related to the exchange of information; and not necessarily commercial,
related to the purchase and sale of products. An obvious example of
this was the case of the photographic equipment manufacturer engaged in
a research programme including technology in which the Soviet Union has
a high degree of expertise; and, to a lesser extent, the National Coal
Board which received useful information on testing procedures.

NOTES

(1) See, for example, the example cited in Hill (1978), pp.209-211.
(2) See Theriot, (1976), pp.734-766. The two reports relating to the
    motives of US companies, quoted by Theriot are:-
    (a) 'Proceedings of the 1974 US Department of Commerce East-West
        Technological Trade Symposium'
    (b) 'National Science Foundation (NSF) - Bureau of East-West Trade
        Survey of 230 major US firms'.
(3) See Theriot (1976), pp.750-751, quoting the Proceedings of the 1974
    US Department of Commerce East-West Technological Trade Symposium.
(4) *Ibid.*
(5) *Ibid.*
(6) See Zaleski, E., 'Central Planning of Research and Development in
    the Soviet Union' in Zaleski, E., et al. (1969), pp.37-127,
    especially the descriptive section of the State Committee for Science
    and Technology (pp.52-68).
(7) See Theriot (1976), pp.750-751, quoting from the 1973 NSF - Bureau
    of East-West Trade survey of 230 major US firms.
(8) *Ibid.*

(9) See McMillan (1977a), p.1186.

(10) This survey of Business International, *Doing Business with Eastern Europe*, October 1975, and Moscow Narodny Bank Bulletins, was carried out for the author by the International Business Unit, Department of Management Sciences, University of Manchester Institute of Science and Technology.

(11) See Therot (1976), pp.763-766.

(12) See Hill (1978), pp.97-101, 209-211.

(13) See Butslov *et al.* (1958).

8 Comments and conclusions, with some international comparisons

INTRODUCTION

This chapter summarises the major conclusions from the research described
in this book, and where possible, relates these to the export performance
of British companies in the markets of the socialist countries of Eastern
Europe. In those cases where information exists, the general export performance of British companies to the socialist countries is also compared
to that of some of their major Western competitors; although such comparisons are necessarily tenuous in view of the small sample size of
companies.

In addition, this chapter also makes recommendations for future studies
in this area of research.

THE PERFORMANCE OF BRITISH EXPORTERS OF CAPITAL GOODS TO THE SOCIALIST COUNTRIES OF EASTERN EUROPE

The final section of Chapter 2 above, revealed that the British share of
Western exports of capital goods to the socialist countries, had generally declined over most of the time interval covered by this research.

It can be hypothesised that this decline would have occurred if those
British firms which initially sold to the socialist markets in the early
and mid-1960s had not been particularly successful in meeting contractual
requirements, causing a lack of customer satisfaction and a low profit-
ability to the seller. If such had been the case, these initial experiences may have had a subsequent snowball effect, discouraging those
same companies from attempting further penetration into the market, and
other British companies from initially entering that market. This
hypothesis is assessed below by reviewing the experiences of British
companies in the Soviet and Eastern European markets.

A study of the experiences of eleven British companies engaged in the
export of capital goods to the socialist countries of Eastern Europe,
has been published previously by the author in 1978 (1), and a further
series of eight case studies of the experiences of British companies
engaged in the export of machine tools and motor industry equipment to
the Soviet Union are contained in Chapter 3 above. The first study was
primarily intended as a means of investigating the marketing methods
employed by successful companies exporting to that region, whilst the

objective of the study included in this present book was to investigate machinery imports as a mechanism of technology transfer into the Soviet Union. They are useful, nevertheless, for the present discussion since some of their results can be compared with information contained in published accounts of the experiences of other Western companies exporting to that region (2).

The results of the research contained in the publications cited above and in Chapter 3 of the present book suggest that those Western companies which have been successful in obtaining, and subsequently fulfilling, contracts with the socialist countries of Eastern Europe have tended to exhibit the characteristics listed below. Most of these characteristics are clearly as relevant to the international marketing of capital goods in general as well as to the socialist countries in particular; although the characteristics listed in paragraphs (d) to (f) below appear to take on added importance when exporting to that region:

(a) the ability to design, manufacture and install products embodying comparatively advanced, but proven, technology, frequently as an integral part of a complex manufacturing system;
(b) the ability to draft technical and commercial proposals in a meticulous manner, over comparatively short time scales;
(c) the ability to co-ordinate the design, manufacture and installation of integrated manufacturing systems for the production of a defined range of products;
(d) the ability to submit repeated but modified versions of these proposals over a lengthy pre-contract time cycle;
(e) the capability to sustain lengthy and highly contended technical and commercial negotiations at the pre-contract stages;
(f) the capability to co-ordinate the timely manufacture, construction and installation of complex plant over quite lengthy time periods, particularly when the volume of orders from the customer may consume a substantial proportion of the company's total resources for that period.

A study of the sources cited above, together with Chapter 3 above, reveals that those British engineering companies which have been successful in export marketing to the socialist markets were generally as capable as their foreign competitors in meeting those requirements listed above, and performing satisfactorily in that market. On the basis of that evidence, therefore, it is this author's view that the hypothesis advanced at the beginning of this section cannot be accepted. In other words, there is no evidence to suggest that those British firms which have sold to the socialist countries have not performed as well as their other Western competitors, in that market.

It is clear, however, that further research could be carried out amongst companies with experience in that market to highlight more clearly the reasons for failure in the market as well as success. This research could well be along the lines of that already carried out amongst Canadian companies by Hannigan and McMillan (3), but with their postal questionnaire method supplemented by structured interviews. Furthermore, it would be useful to carry out similar comparative studies in companies located in other Western countries, particularly Western Europe.

COUNTERPURCHASE

A study of the sources cited in the previous section of this chapter suggests that the willingness to purchase articles manufactured in the socialist countries is an important factor influencing success in that market area. Published statistics show, however, that the rate of increase of UK imports from the socialist countries has been slower than for many other major Western countries (4); which suggests at first sight that British companies may have been generally less willing than most of their Western European competitors to enter into counterpurchase arrangements.

If this has been the case, the lack of willingness to counterpurchase could have been caused by a number of factors, including the comparative sluggishness of the British economy which might have made subsequent distribution more difficult; and the wariness of British companies to counterpurchase certain products, in case they laid themselves open to charges of employment substitution. Clearly the product variety, quality levels and delivery performance provided by Eastern European sellers would also have been important factors discouraging counterpurchase by British companies, although similar problems would have been encountered by their counterparts in other Western countries (5).

On the other hand, although the rate of increase of imports of products made in the socialist markets has been slower for the UK than for many other Western countries, the level of such British imports has still been comparatively high in relation to British exports, particularly imports from the Soviet Union (see Chapter 2 above). These comparatively high recorded values of Soviet imports by the UK, however, may have been influenced by the Soviet use of the London commodity markets to export primary and semi-processed products.

Clearly all of the topics mentioned in this section of the chapter require further investigation, since several cases in this book have shown that the volume and mix of counterpurchases can be crucial factors influencing market success in the socialist countries of Eastern Europe. The most telling examples are those of British Aerospace and Massey-Ferguson-Perkins (see Chapter 5), where counterpurchases were viewed as an important element in those companies' marketing mix when exporting to Romania and Poland, respectively. Furthermore, in the case of British Aerospace, a specific type of business arrangement ('framework contract') was developed for that purpose.

THE ROLE OF PROJECT LEADERS

Chapter 5 above has provided three examples of the important roles played by project leaders in technology transfer and capital equipment exports to the socialist countries of Eastern Europe. It is apparent that the willingness of British Aerospace to assume the status of project leader in relation to the exports of their suppliers' items, as well as the company's own products and technology, was a significant factor in the securing of the relevant orders; and the subsequent high values of market share accounted for by British aircraft exports to Romania. Furthermore, as mentioned in the previous section of this chapter, the company's success in that market was also significantly influenced by its leadership in the formation of framework contracts, to co-ordinate the counterpurchase activities of various principals. Similarly, the comparatively high levels of market share for Polish machine tool imports accounted for by British companies, coincided with the activities of Massey-Ferguson-Perkins as project leader in technology transfer to the Polish tractor and diesel engine industries.

On the other hand, compared with some of their major Western European competitors, British automobile companies appear to have been generally unsuccessful at securing contracts as project leaders for large factory installations in the socialist countries (6); except for some of the automotive component suppliers described in Chapter 4 above, and others referred to by Gutman (7). This absence of British automotive project leaders may have had a detrimental effect on the volume of exports of British-manufactured motor industry equipment to the socialist countries in general, and to the USSR in particular, during a time when the USSR was the world's largest single importer of machine tools (8).

Since the automotive industry has been the second largest single re-corded user of machine tools in the USSR after 'equipment repair' (9), and since a large proportion of Soviet machine tools from the UK, France and Western Germany also consisted of machine tools intended for use in the motor industry (10), it is likely that the majority of Western machine tools imported by the USSR were destined for installation in automobile and associated supplier factories. Published data (see Table 8.1) also reveals that the French and Italian machine tool industries secured a larger market share of Western exports of machine tools to the USSR than their share of total Western machine tool exports, during the time that French and Italian automotive companies were particularly active as co-ordinators of large scale projects in the Soviet Union (11). The large market share held by Western Germany can be explained by its large size as a machine tool builder holding about 50 per cent of the world machine tool export market, and the attention paid to the Soviet market by West German machine tool companies (12). Similarly, the Japanese and American market shares appear to have been increased through concentrated marketing (13). The British market share, however, was one of general decline from 1969 to 1975.

INDUSTRIAL CO-OPERATION

The research described in Chapter 4 above has provided useful information on the types of industrial co-operation agreement entered into between British companies, and organisations in the socialist countries of Eastern Europe; the motives for these agreements; and the extent to which the anticipated advantages were met in practice. For the sample of the thirteen industrial co-operation agreements signed by the ten British engineering companies which were surveyed, it is possible to calculate the frequency with which certain elements of industrial co-operation agreements were included, namely:

managerial services	- 0 cases
capital equipment sale	- 1 case (8% of sample)
complete equipment sale	- 0 case
custom design of plant/equipment	- 1 case (8% of sample)
training of Eastern European personnel	- 6 cases (46% of sample)
technical assistance	- 9 cases (70% of sample)
licence sale	- 10 cases (77% of sample)

Table 8.1

Western Exports of Machine Tools to the USSR (1969-78)
Millions of $US FOB
(Markets Shares shown in brackets)

	1969	1970	1971	1972	1973
UK	27.5(13%)	21.4(14%)	14.9(15%)	22.2(10%)	21.5(7%)
France	26.5(13%)	18.0(12%)	9.1(9%)	15.5(7%)	19.8(6%)
Italy	54.2(26%)	31.7(21%)	10.4(10%)	16.3(8%)	24.5(8%)
FRG	84.0(40%)	62.8(41%)	44.7(44%)	106.8(50%)	168.1(55%)
USA	15.5(7%)	6.2(4%)	13.6(13%)	19.9(9%)	30.5(10%)
Japan	4.2(2%)	12.7(8%)	9.1(9%)	32.1(15%)	42.4(14%)
Total	211.9(100%)	152.8(100%)	101.8(100%)	212.7(100%)	306.8(100%)

	1974	1975	1976	1977	1978
UK	8.7(2%)	12.9(3%)	13.0(3%)	8.6(2%)	8.5(2%)
France	32.0(9%)	52.7(12%)	24.9(5%)	20.7(4%)	42.4(8%)
Italy	39.9(11%)	43.6(10%)	38.8(9%)	42.6(8%)	76.7(15%)
FRG	185.2(50%)	224.0(50%)	307.1(67%)	319.6(64%)	301.3(57%)
USA	68.1(18%)	89.1(20%)	48.1(11%)	31.4(6%)	16.7(3%)
Japan	38.5(10%)	29.0(6%)	23.9(5%)	80.4(16%)	78.5(15%)
Total	372.4(100%)	451.3(100%)	455.8(100%)	503.3(100%)	524.4(100%)

Source: Economic Commission for Europe; *Bulletin of Statistics on World Trade in Engineering Products*; published annually.

supply of parts to Eastern Europe	— 9 cases (70% of sample)
purchase of Eastern components to Western specifications	— 5 cases (38% of sample)
purchase of Eastern finished products to Western specifications	— 1 case (8% of sample)
specialised production and exchange of components	— 0 cases
specialised production and exchange of finished products	— 0 cases
quality control	— 3 cases (24% of sample)
co-ordination of marketing	— 7 cases (54% of sample)
project in a third country	— 1 case (8% of sample)
joint research and development	— 2 cases (16% of sample)

It can be concluded therefore, that a typical industrial co-operation agreement between a British engineering company and a socialist foreign trade organisation has usually consisted of the sale of a licence with the provision of associated know-how, together with the sale of relevant critical components, especially during the start-up phase. Part of the value of the sale to the socialist country has frequently been compensated for by the purchase of related components; but these were usually arranged through separate, although related, contracts. Furthermore, there was frequently a marketing agreement to protect the licensor from unwanted competition, and attempts were made to carry out quality control although on-site difficulties usually reduced this to a pass/fail decision at the component purchase stage.

The main motives for the sample of companies entering into the industrial co-operation agreement were the following:-

(a) income from the sale of a licence, but particularly from the sale of associated components;
(b) possibilities for cheaper sourcing, particularly of less-sophisticated or labour-intensive components;
(c) the use of industrial co-operation as a vehicle for keeping a presence in the Eastern European market, and maintaining a good working relationship with the relevant Eastern European foreign trade organisation.

These motives were also apparent in two of the larger companies described in Chapter 5 (British Aerospace and Massey-Ferguson-Perkins Ltd.) although British Aerospace was also clearly interested initially in the export sales of finished products (aircraft). The third company described in Chapter 5 (GKN Contractors Ltd.) was more interested in the sale of capital goods and technical services related to the technology being transferred.

In the majority of cases, these anticipated advantages which motivated the signing of the industrial co-operation agreement from the British side were borne out in practice, although certain problems were usually encountered, namely:-

(a) inconsistent product quality from the Eastern European partner, especially during start-up;
(b) a lack of breadth in the range of products offered for counter-purchase by the socialist foreign trade organisation;
(c) inconsistencies in the meeting of delivery schedules by the Eastern

European partner;

(d) cumbersome business relationships caused by the bureaucratic methods of industrial management in Eastern Europe.

It is possible to compare these results for the sample of British companies described in this research, with conclusions reported for research on the same topic in other Western countries. One such study was a survey of 218 Western firms carried out by researchers from the East-West Project at the Institute of Soviet and East European Studies, Carleton University, Ottawa. The sample of companies were provided with a questionnaire, from which it was possible to estimate the frequency with which certain types of industrial co-operation activity occurred. This research has been described by McMillan (14), and the major results concerning frequency of various types of industrial co-operation activity are listed below:-

managerial services	8.7% of sample
capital equipment sale	28.4% of sample
complete plant sale	20.2% of sample
custom design of plant/equipment	22.9% of sample
training of East European personnel	46.8% of sample
technical assistance (know-how)	60.1% of sample
licence sale	47.2% of sample
supply of parts to the Eastern partner	53.6% of sample
purchase of East European components to Western specifications	46.8% of sample
purchase of East European finished products to Western specifications	39.9% of sample
specialised production and exchange of components	19.3% of sample
specialised production and exchange of products	5.5% of sample
quality control	25.2% of sample
co-ordination of marketing	31.2% of sample
project in a third country	24.3% of sample
joint research and development	23.9% of sample

Comparing these results with those of the sample of British companies referred to above, therefore, it appears that British companies have not been involved to any great extent in the sale of plant as a consequence of their co-operation agreements, nor joint research and development and activities in third markets. This may be partially explained, however, by the very small sample of British companies compared to McMillan's larger survey, and the restriction of the British sample to engineering companies with their product/component-based technology; whereas McMillan's survey also included chemical companies, many of which would have been engaged in the transfer of process/plant-based technology. Furthermore, the survey described in Chapter 4 above did not include industrial co-operation with the USSR, whereas McMillan's survey did, which may also tend to increase the differences in results between the Carleton survey and the results from the research described in Chapter 4.

As a final point, it is pertinent to compare together the results for co-production in McMillan's survey and the present author's research, with those of the Economic Commission for Europe (ECE) surveys (ECE 1976b and ECE 1978a) referred to in that section of Chapter 2 above which deals with industrial co-operation. Co-production based on the specialisation

of partners was not taking place in any of the British engineering
companies reported in Chapter 4 above, and only in 24.8 per cent (19.3
per cent plus 5.5 per cent) of the returns to McMillan's survey. The
above-cited ECE reports, however, claim that co-production based on the
specialisation of partners accounted for some 30 per cent to 45 per cent
of industrial co-operation agreements, and that this type of arrangement
was more suited to the engineering industry.

The variations noted from the information in the above surveys are
probably due to the interpretation of the term 'co-production based on
the specialisation of partners'; which the present author considers to
mean a carefully planned and executed schedule of specialised production
and subsequent delivery of components, or finished products, to meet both
partners' manufacturing plans. In many of the cases covered in Chapter
4 above, however, the socialist partner usually only purchased those
components which he was incapable of producing until the technology was
fully assimilated; and the British partner usually only purchased
components when the Eastern European products were significantly cheaper,
or during shortfalls in domestic production capacity, or to meet counterpurchase requirements. It is possible, however, that other researchers
may have interpreted the two-way delivery of components, however
intermittently planned or loosely scheduled, as evidence of co-production
based on the specialisation of partners.

In addition to their postal survey, researchers from Carleton
subsequently visited a sub-sample of fourteen Western European engineering companies in 1974 and 1975, for interviews with industrial
executives. This sample of companies was made up of the following:-

   a West German manufacturer of construction machinery,
   a West German manufacturer of machine tools,
   a West German manufacturer of electrical products,
   a West German manufacturer of trucks and buses,
   an Italian automobile company,
   a French manufacturer of marine diesel engines,
   a French computer manufacturer,
   an Austrian manufacturer of trucks and buses,
   an Austrian office machinery manufacturer,
   a Swedish manufacturer of system control components,
   a Swedish manufacturer of submersible pumps,
   a Swedish manufacturer of cars and trucks,
   a Swedish bus body-builder,
   a Swedish computer manufacturer.

Although the interviews with these companies were less structured in
format than those reported by the present author, it is still possible
to gain much useful information from the interviewers' notes which were
made available to the author when he visited Carleton during 1980.

The topics mentioned as possible motives for industrial co-operation,
and the frequency with which they were referred to by the sample of
companies are listed below:-

   increased market opportunities   - 11 companies
   cost savings from East European-based
      production   - 7 companies
   speedier technological absorption within

```
the company                                -    2 companies
stability of production, and technical
    capacities, of the socialist partner  -    no company
opportunities provided for specialised
    production of certain products and
    components                             -    2 companies
other technical spin-offs                  -    no company
preferential customs treatment
    provided for items manufactured under
    the co-operation agreement             -    1 company
reduced transport costs                    -    no company
```

Since most of the motives for industrial co-operation of this sample
were in the areas of 'increased market opportunities' and 'cost savings'
it was decided to report these in more detail in summarised case study
style, as shown in Table 8.2, together with other advantages and disadvantages from the operation of the agreement. From this table, it
appears that at least eight of the fourteen companies were benefiting
from the sale of components to Eastern Europe through the receipt of
income which could not otherwise have been obtained because of hard
currency restrictions. In addition, seven of the companies were gaining
income from licence fees, four of which were related to the sale of
components. Eight of the companies also expected to obtain certain cost
savings from sourcing in Eastern Europe, sometimes as a consequence of
cheaper labour costs compared with Western Europe, but also frequently as
a consequence of economies of scale from increased volumes of output.
These cost differences were thought to be crucial to product sales in at
least two cases: in the first, the market had become increasingly price
competitive; and in the second, the Western market volumes were decreasing as a consequence of maturity in the product life cycle. Furthermore,
in these latter cases, the Eastern European partner was considered as a
critical source for spares. There were few other advantages listed apart
from increased chances of entry into other socialist markets, and tariff
advantages in certain circumstances; although one company (a West German
machine tool manufacturer) did mention the positive role played by its
Hungarian partner in carrying out product technical development.

    Clearly, it is difficult to directly compare the information obtained
from the British and Western European companies in view of the smallness
of the sample sizes, and the lack of use of a structured format in the
Western European sample. It can be asserted, however, that a large
proportion of both samples of companies entered the co-operation arrangements to increase their sales of licences, components, or finished
products, and to increase their presence in the Eastern European market.
On the other hand, it appears that there may have been differences in
attitude concerning counterpurchases of Eastern European products
between the two samples of companies; several of the Western European
firms appeared to view the Eastern partner as a cost-competitive reser-
voir of production capacity, particularly for finished products, whilst
very few of the British firms appeared to hold that opinion with the
same degree of enthusiasm. It is important to note, however, that the
survey of the West European companies was carried out some five years
before the survey of their British counterparts; and consequently some
of their expectations regarding Eastern European capacities, may not
have been realised.

    The sets of problems faced by companies engaged in industrial co-operation appeared to be almost identical in both samples, namely inconsistent product qualities and delivery reliabilities, and slow decision-making and action as a consequence of the bureaucratic systems of management in the socialist countries. Similar problems were also encountered by a larger sample of Western European companies reported by Levcik and Stankovsky (5).

Further comparisons may also be made from information available on five Western companies that have signed industrial co-operation agreements with the same Eastern European foreign trade enterprise, and associated manufacturing activities carried out in the same Eastern European factory. The sample of Western companies is made up as follows:-

(a) two British crane manufacturers, one of which is described in Case Study No. 1 of Chapter 4 above, whilst the other is referred to in Paliwoda (6);
(b) a West German manufacturer of concrete mixers and concrete pumps (see Appendix E for an account of this unpublished case);
(c) an American manufacturer of agricultural and handling machinery (7);
(d) an American manufacturer of transmission units (8).

Each of the above companies were involved in industrial co-operation activities with the same Polish foreign trade enterprise, responsible for the import and export of construction machinery, and all relevant production activities were apparently carried out by the same Polish factory.

The information available from these studies leads to the conclusion that this small sample of British, American and West German companies became involved in industrial co-operation with the Polish partner primarily to sell components, and improve their working relationships with the relevant foreign trade organisation. In addition, there were believed to be useful opportunities for cheaper sourcing, particularly for less sophisticated components. Furthermore, the problems encountered in industrial co-operation by the British companies were not substantially different from those encountered in the non-British sample, namely: inconsistencies in Polish product quality, especially during start-up; a narrow range of products offered for counterpurchase; inconsistencies in the meeting of delivery schedules by the Polish partner, and cumbersome business relationships caused by the bureaucratic methods of industrial management in Eastern Europe.

It is clear therefore, that there is still further scope for research on industrial co-operation using a larger sample of companies, and a standardised methodology, in order to enable inter-country comparisons to be made between companies from different Western states. Furthermore, the use of a larger sample may also enable more information to be obtained on industrial co-operation arrangements with foreign trade organisations from the less-represented countries in the sample in this book, namely Bulgaria and Czechoslovakia; but particularly from the GDR, for which no single case of industrial co-operation or technology transfer could be found with a British company. Such cases are still likely to be small in number, however, since the results of this research on British companies have reflected the results of the previous studies quoted in Chapter 2 above, namely that Hungary, Poland and Romania are the predominant socialist countries engaged in industrial co-operation with Western companies apart from the USSR.

Table 8.2

Features of Industrial Co-operation Agreements between Socialist Foreign Trade Organisations and Fourteen Western European Companies

Type of Company	Details of Co-operation Agreement	Expansion of Market Opportunities	Expected Cost Savings	Other Advantages	Disadvantages
A West German manufacturer of construction machinery (see Appendix E for further details)	Co-operation with a Polish foreign trade organisation and enterprise for the manufacture of mobile concrete mixers and cement pumps (signed in 1970). The West German company supplies hydraulic components, whilst the Polish partner carries out the fabrication (about 60% of total product value).	Export of components to Poland	Cost savings in labour intensive fabrications	Tariff reductions for components to be subsequently re-exported in a manufactured product.	Some problems have been encountered in quality, delivery and co-ordination between the foreign trade enterprise and the factory.
A West German manufacturer of milling machines, optical grinders and other machine tools.	Co-operation with a Hungarian foreign trade organisation and enterprise for the assembly of milling machines, using some West	Possibility of export into other markets in the socialist countries (e.g. machines for the Kama River project).	Cheaper assembly costs in Hungary thus making the machines more competitive in Western markets.	(a) Good quality of Hungarian production. (b) Hungarian partner has the technical capability to contribute relevant	Annual commitment to purchase a prescribed quantity of machines.

Company	Co-operation activities	Market rights / Income	Cost savings		Problems
	German produced components.	Each partner has exclusive rights in their own regional markets, although each can sell in the other's markets with permission.		know-how and expertise. (c) Royalties received for Western sales.	
A West German manufacturer of a wide range of electrical engineering products.	The company has signed licensing or industrial co-operation agreements with the majority of the East European states. These include: (a) manufacture of electrical grills for domestic cookers in Hungary; (b) manufacture of domestic electrical irons in Poland; (c) licence sale to USSR for manufacture of magnetic filters used in boilers; (d) purchase of telephone exchange components from Bulgaria;	Income from royalty payments, and some sales of components.	In certain cases, where cost savings can be made through economies of scale for the manufacture of some components.	None.	(a) Differences in technical standards of operation between the company and the socialist countries. (b) Differences in attitudes towards marketing and deliveries, between the company and the socialist countries. (c) Some problems in managerial control, even in joint ventures.

Table 8.2 (continued)

Type of Company	Details of Co-operation Agreement	Expansion of Market Opportunities	Expected Cost Savings	Other Advantages	Disadvantages
	(e)purchase of tape punches from Czechoslovakia; (f)joint product-ion of computer software with Hungary.				
A West German manufacturer of trucks and buses.	(a)A ten year agreement was signed in January 1969 with a Romanian foreign trade organisation, for the manu-facturing rights of certain models, and the sale of these in the CMEA areas and China. Some of the models are now obsolete. (b)An agreement was signed with a Hungarian foreign trade organisation for the manufacture of engines. In 1971, this was extended to	(a)Income from licence fees. (b)Income from product sales until capacity is assimilated. (c)Income from component sales until capacity is assimilated.	Since Eastern European labour costs are lower than those in Western Germany, it may be more economical to source spares from the social-ist partner, particularly when the model becomes obsolete.		(a)Obligation to counterpurchase from Romania. (b)Problems of product quality, and of delivery reliability of Romanian partner. (c)Possibility of future competition from Hungarian partner.

172

	include a limited range of truck models.		
An Italian automobile company.	(a)Sale to the USSR of a licence to manufacture a sturdier variation of one of its models, together with manufacturing know-how. (b)Similar licence and manufacturing know-how sales to a Polish foreign trade organisation.	(a)Income from licensing and know-how fees. (b)Possibilities of future business contacts.	Competition in Western markets, except Italy.
A French manufacturer of marine diesel engines.	A ten-year co-operation programme related to the sale of a marine engine licence, to a Soviet ministry.	(a)Sale of technical documentation. (b)Sale of components during the assimilation period. (c)Royalty payments related to horsepower of engines manufactured. (d)Possibilities of further sales to other East European countries.	(a)Possibility of purchase of Soviet-produced crankshafts. (b)Possibility of a purchase of a Soviet licence for turbo-blowers.

Table 8.2 (continued)

Type of Company	Details of Co-operation Agreement	Expansion of Market Opportunities	Expected Cost Savings	Other Advantages	Disadvantages
A French manufacturer of computers (government-owned).	(a)Sale of licence to produce a mini-computer, to a Hungarian foreign trade organisation. (b)Sale of a licence and manufacturing know-how for a series of computers, to Romania.	"Pays for itself".			(a)The company is comparatively small and does not have much demand for some of the products offered in counter-purchase. (b)COCOM has some-times objected to some of the sales. (c)A hope for in-creased sales has not materialised.
An Austrian manufacturer of trucks and buses.	(a)Originally licensed manu-facture of eng-ines, gearboxes and clutches to Hungary in 1947. (b)An agreement for the joint manufacture of heavy agricult-ural tractors in Hungary in 1968. (c)An agreement for the joint	Sale of components.	Cost savings from components manufactured in Eastern Europe.		(a)The Hungarian partner almost uni-laterally ceased tractor production as a consequence of a CMEA decision. This could have caused problems in sourcing for spares. (b)Quality problems with some Hungarian -produced components for certain models of trailer. (c)The rate of

174

	manufacture of trailers in Hungary in 1969. (d)An agreement for the manufacture of buses in Hungary in 1970 (e)An agreement has also been signed with Poland on counterpurchase of Polski Fiats in exchange for Austrian trucks; and joint production of trucks based on Polish engines and Austrian chassis.			exchange applied by the Hungarian authorities made the Austrian-produced components too expensive for the Hungarian purchaser. (d)The market volume for the finished product in the West was too low to generate a high volume of counterpurchases.
An Austrian manufacturer of office machinery.	The company has signed co-operation agreements with foreign trade organisations and production enterprises in Hungary and Czechoslovakia for the joint production of cash registers. Shared production of components have been	(a)Sale of components into Eastern Europe. (b)Able to maintain a competitive position in shrinking Western markets for mechanical cash registers.	Lower production costs, partly through lower, and more stable unit costs in Eastern Europe, but also as a consequence of longer production runs in each partner's factory.	Initial problems of Eastern European product quality, but these were soon overcome.

Table 8.2 (continued)

Type of Company	Details of Co-operation Agreement	Expansion of Market Opportunities	Expected Cost Savings	Other Advantages	Disadvantages
	arranged on standard minute values, and components exchanged for specialised final assembly in each of the East European partners' factories. The company has also sold a licence to Poland for the production of franking machines.				
A Swedish manufacturer of hydraulic pneumatic and electrical components for automation systems.	Licensed production of a range of items, to a Hungarian enterprise with foreign trade rights. Counter-purchase of some items made under licence. New components sent to Sweden for testing and approval.	Royalty payments on Hungarian production up to an agreed maximum figure.	Components provided from Hungarian partner at discount compared to the Swedish costs.		(a) Some initial promotion necessary in Eastern Europe for acceptance of Hungarian-made components. (b) Initial quality problems encountered amongst some suppliers to Hungarian partner.

A Swedish manufacturer of sub-mersible pumps.	Licensed production of a range of pumps to a foreign trade enterprise and production organisation in Hungary. Some purchase of Hungarian components by the Swedish company. Sales restricted to Hungary, although some export possible when sold as part of a system.	Income from sale of components.	A good entry into Hungarian market, licenced products account for a large share of the market.	None.
A Swedish manufacturer of cars and trucks.	Joint company with Hungarian foreign trade organisations for distribution of a special vehicle in the Hungarian market; the vehicle is sold elsewhere by the Swedish company. The vehicle is assembled in Hungary using engines and gear-boxes purchased from the Swedish	Sale of product in Hungarian and Western markets, under conditions of production capacity con-straints in Sweden.	The volumes in-volved in both markets now makes production economic again, following a fall in demand in Western markets.	(a) Delays in launch of production. (b) Business oper-ations sometimes over-formal.

177

Table 8.2 (continued)

Type of Company	Details of Co-operation Agreement	Expansion of Market Opportunities	Expected Cost Savings	Other Advantages	Disadvantages
	company, and rear axles from elsewhere.				
A Swedish bus body-builder.	Co-operation with Hungarian foreign trade organisations for Hungarian production of bus bodywork parts, and final assembly of buses using these parts and Swedish items. Buses sold in Western markets.		Production costs make manufacture economic.		Very formal in business operation.
A Swedish manufacturer of computers.	Sale of licence to a Hungarian foreign trade organisation and manufacturing enterprise for the production of central processing units and peripherals. Counterpurchase of these and other products. There are no marketing restrictions.	Sale into the Hungarian market.		Alternative source for central processing units.	

As a final point, it is this author's view, that the management of counterpurchases, particularly within the context of industrial co-operation agreements, requires further research.

THE INFLUENCE OF INTER-GOVERNMENTAL RELATIONS

Official inter-governmental political relations are frequently considered to play a major role in the level of business between the socialist countries of Eastern Europe and the Western states. In the highly centralised and bureaucratic Eastern European markets, official approval and support from relevant central organs, as a result of amicable official relations, may frequently assist a Western company in its dealings with associated foreign trade organisations.

As explained in Chapter 1 above, the British government was probably the first western government to sign a long term trade agreement with an Eastern European socialist country during the post-war years, namely the 1959 Anglo-Soviet Trade Agreement (19). The UK also signed a series of inter-governmental trade and technological co-operation agreements with the socialist countries during the 1960s and early 1970s, culminating in a series of agreements of economic, scientific and technological co-operation, during the mid 1970s; many of which gave lists of promising areas for future co-operation. The progress of these agreements has usually been monitored annually by senior trade officials meeting in a Joint Commission and reports published accordingly, some of which have included additional, or more closely defined, industrial topics for further co-operation.

On the strength of these inter-governmental arrangements, therefore, it would appear that Britain's trade performance in Eastern Europe cannot be ascribed to poor official relations. Furthermore, the majority of British financial institutions appear to have been prepared to provide competitive credit facilities to the socialist countries for the purchase of British products, particularly capital goods, and many of these facilities have been officially supported by the British government through its Export Credit Guarantee Department (20).

When Britain's export trade in engineering products to the USSR is viewed in more detail, however, it is clear that British performance in that market was worse than in the remaining socialist markets (21) during the early and mid-1970s. In this author's view, that position was partly caused by political factors, namely the expulsion of some one hundred Soviet diplomats and trade officials from London in 1971, for alleged activities detrimental to constructive inter-governmental relations. As a consequence of Soviet retaliation to that action, many British companies probably lost the close contact with Soviet foreign trade organisations which they had painstakingly built up over a number of years. In some instances, this loss of contact would have occurred at crucial stages of tendering and contract drafting, and the consequent potential orders lost to foreign competitors (22).

The Anglo-Soviet inter-governmental agreement of 1974, and the subsequent 1975 long term programme for economic, industrial, scientific and technological co-operation attempted to 'normalise' political and trading relationships between the two countries. This programme gave detailed lists of the manufacturing sectors where industrial co-operation was potentially most promising, and also classified them in detail in terms of the types of trade and industrial co-operation to be carried out (e.g. supply of production equipment, supply of plant through turn-key arrangements etc.); furthermore, the programme was also supported with a generous credit facility from the British government. This worth-while attempt at improvement at inter-governmental relations apparently went some way to reverse the trend set during 1971 to 1974, when access to Soviet foreign trade organisations was probably more difficult for many British companies. Data shown above in Tables 2.2 and 2.10, for example, shows that from 1974 to 1978, British exports to the USSR trebled, compared to a doubling in exports to the USSR by the developed market economies in total; although the British share still remained at less than 5 per cent of the developed market economies total exports to the Soviet Union.

The argument put forward in the previous two paragraphs is also further supported by evidence provided by information in the case studies con-tained in Chapter 5 above. It is apparent from that chapter that the support of the British government for British Aerospace's and Massey-Ferguson-Perkins' industrial co-operation arrangements with Romania and Poland respectively, were positive factors influencing the companies' successes in those markets.

It is clear, therefore, that the role played by government in East-West trade, technology transfer and industrial co-operation is one that is worthy of further research. This should attempt to directly relate the progress of inter-governmental relations to the progress of trade nego-tiations related to specific contracts, and also to explore in more detail the working practices that have evolved for integrating govern-ment policy and business strategy for trade with this politically sensitive market. Such research would consequently cover the two important facets of the role of government in East-West trade, namely external political relations with the socialist countries of Eastern Europe; and internal industrial and commercial relationships with comp-anies engaged in East-West trade and technology transfer.

GENERAL PERFORMANCE IN INTERNATIONAL MARKETS

In Chapter 2 above it has been shown that the share of Western goods delivered by Britain to the socialist countries of Eastern Europe, has been in general decline since the early 1960s. This has been particul-arly apparent for engineering goods where the UK appears to have lost its market share to its traditional Western European competitors, and also to the USA - a comparative newcomer to this market.

At the same time, however, it is important to note that British export performance to the world as a whole has been in comparative decline. In 1963, Britain accounted for 11.5 per cent of the developed market economies' total exports, but this had fallen to almost 9 per cent in 1970 and 8 per cent in 1977 (23). For engineering goods, Britain's market share of total world exports from the developed market economies had fallen from 16 per cent in 1963 to 10 per cent in 1970, and to less than 8 per cent by 1977 (24). Similarly, for 'machinery non-electric', Britain accounted for 21 per cent of the value of the total world exports of UK, France, Italy, FRG, USA and Japan in 1963, but this share had fallen to 15 per cent by 1970 and 12 per cent by 1977 (25).

Furthermore, data plotted in Figure 8.1 below demonstrates that for the selected Western countries, the market share of machinery exported by any one Western country to the socialist countries, is not dissimilar from the machinery share held by that country in the total world export market (26). This further suggests, therefore, that there are certain common factors influencing a country's export performance in the world markets in general, which also influence that country's performance in the socialist countries in particular. These former factors probably include product promotion, product development, sales organisation, pricing, delivery and after-sales service (27), and changing government national-economic and international policies; in addition to those factors which appear to be specific to exporting to the socialist countries discussed in previous sections of this chapter. It is important, therefore, that any further research on British companies export performance to the socialist markets also includes an element which allows for their performance in international markets in general.

The USA has clearly been an exception to this general rule discussed in the above paragraph (28), as a result of poor inter-governmental relations: that country had been slow to establish itself in the export markets of the socialist countries prior to the signing of an inter-governmental trade agreement with the USSR in 1972 (29), although improved performance of US companies in the Soviet engineering market occurred in the early 1970s, chiefly as a consequence of American machinery exports for the Soviet truck, metallurgical and oil industries (30), and the availability of US government credit support. The US market share of non-grain exports was not maintained, however, coincident with a corresponding deterioration in US/Soviet inter-governmental relations.

MARKET PREFERENCES OF BRITISH COMPANIES

From published statistics on the UK export of engineering goods it is apparent that the largest single market region for British engineering products has been Western Europe, which purchased 38 per cent of British engineering exports in 1964, 43 per cent in 1970 and 40 per cent in 1975 (see Table 8.3). From this source it is also apparent that the 'English speaking industrially developed nations' (i.e. USA, Canada, Australia and South Africa) could be considered as the second largest market, accounting for 27 per cent of British engineering exports in 1965, 25 per cent in 1970 and 21 per cent in 1975 (see Table 8.4).

The apparent preferences for British companies to sell to those markets has been conditioned by a series of economic, political, historical and geographical factors, namely:

(a) the large size of the Western European and English-speaking markets;
(b) the geographical proximity of the large Western European markets;
(c) the historical British trading links with Australia, Canada and South Africa as a consequence of present and previous Commonwealth ties;
(d) the previously common ties of language and measurement systems with the English-speaking countries;
(e) the friendly and comparatively stable political relations enjoyed between the governments of the UK, and those of the English-speaking countries and Western Europe, since 1945.

181

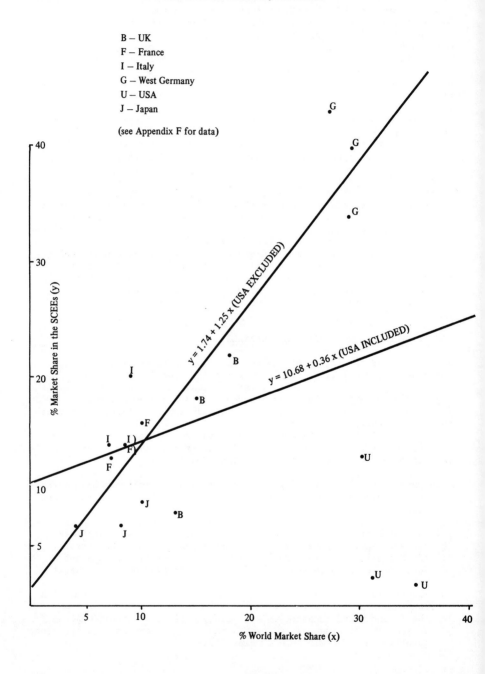

Figure 8.1

Market Shares for "Machinery Non-Electric"

B – UK
F – France
I – Italy
G – West Germany
U – USA
J – Japan

(see Appendix F for data)

$y = 1.74 + 1.25 x$ (USA EXCLUDED)

$y = 10.68 + 0.36 x$ (USA INCLUDED)

% Market Share in the SCEEs (y)

% World Market Share (x)

Table 8.3

Western Engineering Exports to Western Europe and Eastern Europe
Deliveries shown in millions of $US FOB
(% of deliveries shown in brackets for each country)

From	UK	France	Italy	FRG	USA	Japan
To			1964			
W. Europe	1936.1 (37.8%)	1291.7 (56.1%)	1154.7 (62.5%)	4918.2 (65.3%)	2069.3 (22.1%)	2549.0 (13.0%)
SCEES	119.7 (2.3%)	81.6 (3.5%)	86.8 (4.7%)	261.6 (3.5%)	7.5 (8.1%)	131.5 (6.7%)
Total	5112.5 (100%)	2302.4 (100%)	1846.2 (100%)	7531.3 (100%)	9350.3 (100%)	1958.0 (100%)
			1970			
W. Europe	3368.6 (42.5%)	3555.2 (60.6%)	3148.0 (64.6%)	10297.6 (64.8%)	5302.5 (29.7%)	1326.8 (16.9%)
SCEES	248.1 (3.1%)	310.6 (5.3%)	323.0 (6.6%)	482.7 (3.0%)	69.0 (0.4%)	125.6 (1.6%)
Total	7922.9 (100%)	5864.9 (100%)	4867.4 (100%)	15898.1 (100%)	17881.9 (100%)	7833.8 (100%)
			1975			
W. Europe	7325.3 (40.1%)	10176.3 (53.8%)	6609.0 (54.2%)	24303.6 (58.1%)	10458.4 (23.0%)	4305.5 (16.0%)
SCEES	573.3 (3.1%)	1259.4 (6.7%)	844.1 (6.9%)	2783.7 (6.7%)	767.2 (1.7%)	835.4 (3.0%)
Total	18236.4 (100%)	18900.3 (100%)	12195.3 (100%)	41855.9 (100%)	45709.4 (100%)	27405.0 (100%)

Source: Abstracted from Economic Commission for Europe *Bulletins of Statistics on World Trade in Engineering Products*, published annually.

Table 8.4

Proportions of Western Engineering Exports to English-speaking
Markets (i.e. USA, Canada, South Africa and Australia)
(USA shown in brackets)

Year	1965	1970	1975
Countries			
UK	27%(8%)	25%(10%)	21%(10%)
France	8%(4%)	8%(4%)	6%(3%)
FRG	14%(10%)	17%(12%)	12%(8%)
USA	32% -	36% -	31% -
Japan	27%(22%)	39%(31%)	30%(22%)

Source: Compiled from export sales data for each country reported in
Economic Commission for Europe, *Bulletin of Statistics of World Trade in
Engineering Products* for each country.

Table 8.5

Imports of Engineering Products by Western Europe and
English-speaking Markets
(all figures in $US m)

	1965	1970	1975
Australia	1151.0(3%)	1641.0(2%)	3829.9(2%)
South Africa	1058.0(2%)	1722.6(2%)	4231.7(2%)
Canada	3093.0(7%)	5833.1(7%)	14740.9(6%)
USA	3280.2(7%)	11428.0(13%)	23437.9(10%)
Western Europe	18126.0(40%)	36511.0(41%)	89847.1(37%)
SCEES	5064.0(11%)	9587.0(11%)	26585.6(11%)
Total World	45085.0	89454.1	240225.7

Source: Economic Commission for Europe *Bulletin of Statistics of World
Trade in Engineering Products*, Table 4. (1965, 1970 and 1975).

Note: Market shares shown in brackets (i.e. for Australia $\frac{1641}{89454.1}$ = 2%
in 1970).
The total world trade figure for 1965 does not include Bulgaria
and Romania.

It can be argued, therefore, that the socialist countries of Eastern Europe presented themselves as almost 'markets of last resort' to the majority of British exporters, compared with the English-speaking and Western European markets referred to above. It is important to consider, however, whether these British 'preferred markets' sustained a comparatively high growth rate for imports, and also whether British exporters succeeded in maintaining their shares of these imports.

With regard to the first of these questions, it is apparent that the country with the highest growth rate in the import of engineering goods was the USA - over 600 per cent between 1965 and 1975, accounting for the receipt of some 10 per cent of the total world exports of engineering goods in 1975 (see Tables 8.5 and 8.6). Western Europe, on the other hand, was the largest single market, (although a large proportion of this region's imports would be delivered by member countries) receiving almost 40 per cent of the world's exports of engineering goods in 1975, but with a slower growth rate than the USA over the 1965 to 1975 period (almost 400 per cent). The socialist countries of Eastern Europe had a slightly higher growth rate than Western Europe for the imports of engineering goods (see Table 8.6) and were similar to the USA in terms of overall size. Finally, Australia, Canada and South Africa together similarly received approximately some 10 per cent of the total world deliveries of engineering products in 1975, but these imports only grew by some 330 per cent from 1965 to 1975.

With regard to the second of the above questions it is also apparent that the British share of the engineering goods markets in Australia, Canada and South Africa was in quite rapid decline, reducing from 36 per cent in 1965 to 18 per cent in 1975 in Australia, 39 per cent in 1965 to 22 per cent in 1975 in South Africa, and 7 per cent in 1965 to 3 per cent in 1975 in Canada (see Table 8.7). Consequently, these major British markets were not only expanding at a slower rate than other market areas, but the proportion of British goods imported by those markets was also in decline. In the United States market, the British market share of imported engineering goods was similarly in decline, from 14 per cent in 1965 to 7 per cent in 1975. In Western Europe, the decline in British market share was less marked, however; but it still appears that the decline of British market shares for engineering goods has been general, even in those areas of the world which have been traditionally viewed as 'British markets' (see Table 8.7).

Different exporting characteristics have been apparent for the UK's competing countries, however. In the first place, Britain's Western European competitors appear to have been more 'European-oriented', delivering a far larger proportion of their engineering exports to both Western and Eastern Europe, than has been the case for the United Kingdom. For example, France, Italy and the Federal Republic of Germany sold between 55 per cent and 65 per cent of their total engineering exports to other Western European nations between 1964 and 1975 (see Table 8.3) and between 3.5 per cent and 7 per cent of their engineering exports to the socialist countries of Eastern Europe. The United Kingdom, on the other hand, sold only about 40 per cent of her engineering exports to the Western European market and between 2 per cent and 3 per cent to the socialist countries of Eastern Europe. Looking at the Eastern European markets in more detail, data published in the *Bulletin of Statistics on World Trade in Engineering Products* for 1977 show that both France and Italy sold almost twice as many engineering goods as the

Table 8.6

Growth Rates for Imports of Engineering Products by
Western European and English-speaking Markets, and the SCEEs

	1965	1970	1975	Calculated Growth Rates 1975/1965
Australia, Canada & S. Africa	5302.0	9197.0	22803.0	330%
USA	3280.0	11428.0	23438.0	615%
W. Europe	18126.0	36511.0	89847.1	395%
SCEES	5064.0	9587.0	26585.0	425%

Source: Calculated from data shown in Table 8.5 above.

Table 8.7

British Market Shares for Engineering Products
(SITC Group 7)

Importers	1965	1970	1975
Australia	36%	25%	18%
S. Africa	39%	27%	22%
Canada	7%	5%	3%
USA	14%	7%	7%
W. Europe	(11%)	9%	8%
E. Europe	(4%)	3%	2%
Total World	12%	9%	8%

Source: Compiled from data shown in Economic Commission for Europe;
Bulletin of Statistics on World Trade in Engineering Products 1975,
Tables 1 & 4, (1965, 1970 and 1975).

Note: Estimates from 1964 and 1966 data shown in brackets.

UK to the socialist countries, but only about 10 per cent more than the
UK into the world as a whole in the case of France, and 30 per cent less
than the UK in the case of Italy. Similarly, in 1977 Western Germany
sold almost six times as many engineering goods as the UK to the socialist countries, but only about two and a half times to the world as a
whole. Furthermore, these Western European countries appear to have
sold a far smaller proportion of their engineering goods to the English-
speaking export markets, with the exception of the Federal Republic of
Germany, which has sold approximately the same proportion of its engineering exports as the UK to the USA (see Table 8.4).

In the case of USA, its proportion of engineering exports delivered to
the socialist countries has been comparatively small, the traditional
American export markets being those of Canada, Japan and to a lesser
extent, Western Europe. In view of its large size and high level of
technological expertise, however, and its growing interest in the socialist markets in the 1970s, particularly for the sale of automotive
industry equipment to the USSR, the USA was successful in increasing its
share of the Eastern European engineering market between 1965 and 1975.
Finally, the proportion of Japanese engineering exports delivered to the
socialist countries has stayed at about the same level: the Japanese
engineering industry, therefore, appears to have expanded its sales in
the Eastern European export markets as a direct consequence of the overall expansion of its engineering industry, and penetration into all
export markets.

Similar patterns are also apparent for exports of 'machinery non-
electric', with France, Italy and Western Germany tending to export a
higher proportion of their goods (in 1975) to Europe in general and to
the socialist countries of Eastern Europe in particular. The United
Kingdom, on the other hand, has tended to export a larger proportion of
capital goods to the 'English-speaking markets' (see Tables 8.8 and 8.9).

It is this author's view, therefore, that any further research on the
export performance of British companies in the socialist markets should
take account of the apparent export preferences of those companies.

THE PURCHASE OF LICENCES FROM THE SOCIALIST COUNTRIES

Chapter 6 above contains a survey of a sample of British companies that
have purchased industrial licences from the socialist countries. The
number of such licence purchases appears to be very small, and the
majority appear to be related to the metallurgical industries. In
general, however, it was found that the purchases of such licences were
entirely satisfactory from the commercial viewpoint.

TRIPARTITE INDUSTRIAL CO-OPERATION

Only one of the cases (the electronics company described in Chapter 7
above) included in this book contained any element of tripartite industrial co-operation, in which a socialist foreign trade organisation
and a Western company co-operate to sell products or technology to a
third (usually developing) country. A sample of these agreements
compiled and studied by Gutman (31) reveal that UK companies probably
accounted for between 5 per cent to 10 per cent of such contracts signed

Table 8.8

Western Exports of 'Machinery non-electric' (SITC Group 71)
(all figures in $ millions FOB)

From:	UK	France	Italy	FRG	USA	Japan
To:			1964			
Australia	164.0	4.0	11.3	40.9	201.3	52.2
Canada	110.1	3.5	7.6	31.5	1139.9	5.6
S. Africa	139.9	12.3	16.4	59.6	124.8	10.7
USA	124.7	21.6	38.5	168.2	-	71.3
W. Europe	985.1	522.6	521.2	2423.0	1375.0	43.0
SCEES	97.8	54.2	44.6	178.9	5.8	52.8
Total World	2402.7	910.8	863.0	3588.1	4718.6	481.3
			1970			
Australia	198.4	13.0	23.8	90.2	288.0	84.0
Canada	155.1	10.6	24.2	90.7	1835.3	52.4
S. Africa	218.1	40.5	59.5	152.6	177.3	40.4
USA	392.1	84.0	167.1	556.7	-	410.4
W. Europe	1694.8	1291.6	1402.7	5002.4	2723.2	309.9
SCEES	198.6	178.1	219.9	380.7	56.8	81.5
Total World	3941.1	2248.5	2499.1	7620.8	8380.1	2006.2
			1975			
Australia	321.1	27.6	51.8	171.3	574.4	172.8
Canada	304.7	52.7	72.4	227.2	4622.8	141.9
S. Africa	453.5	109.9	117.1	408.9	472.9	116.7
USA	960.3	259.1	296.2	1111.6	-	958.0
W. Europe	3794.5	4425.6	2914.2	11607.4	5646.5	986.5
SCEES	406.8	803.3	669.2	1927.8	626.1	459.9
Total World	9420.6	7404.5	6010.6	20277.5	20899.7	6729.7

Sources: Economic Commission for Europe; *Bulletin of Statistics on World Trade in Engineering Products* for 1964, 1970 and 1975.

Table 8.9

Proportions of Exports of 'Machinery non-electric'
(SITC Group 71), for selected Western Countries

From:	UK	France	Italy	FRG	USA	Japan
To:				1964		
'English speaking markets' (USA in brackets)	23% (5%)	3% (2%)	8% (4%)	9% (5%)	31%	29% (15%)
Western Europe	41%	57%	60%	68%	29%	9%
SCEES	5%	6%	5%	5%	0%	11%
				1970		
'English speaking markets' (USA in brackets)	25% (10%)	7% (4%)	11% (7%)	11% (7%)	27%	29% (20%)
Western Europe	43%	57%	56%	66%	32%	15%
SCEES	5%	8%	9%	5%	1%	4%
				1975		
'English speaking markets' (USA in brackets)	21% (10%)	6% (4%)	9% (5%)	9% (5%)	27%	21% (14%)
Western Europe	40%	60%	48%	57%	27%	15%
SCEES	4%	11%	11%	10%	3%	7%

Sources: Data shown in Table 8.8 above.

between 1965 and 1979, compared with 21 per cent and 24 per cent for FRG companies, 15 per cent to 27 per cent for French companies, and 8 per cent to 13 per cent for Italian companies.

It is possible, however, that many of the technologies used by the British engineering companies described in this book do not lend themselves to joint exploitation in a third (developing country), since many of the technologies mentioned here relate to capital equipment, or components for transport equipment; whereas 70 per cent of the tripartite contracts surveyed by Gutman (32) were in the energy or intermediate goods industries. Alternatively, British companies may not have regarded tripartite industrial co-operation as a particularly viable vehicle for export trade, but further research is clearly required to test either of these hypotheses.

TECHNICAL CO-OPERATION AGREEMENTS

The research described in this book has reported the experience of a sample of British companies in the area of technical co-operation agreements with organisations in the socialist countries of Eastern Europe, but chiefly in the Soviet Union (see Chapter 7 above). In general, it was found that these agreements were moderately successful when the objectives of the agreement were clearly defined in terms of the exchange of technical information, but disappointing when a large volume of trade was expected.

IMPACT ON EAST EUROPEAN INDUSTRIES

Several researchers have attempted to study the effect of the purchase of Western technology and equipment on the performance of specific industrial sectors in the socialist countries (33). It is this author's view that in certain instances the information contained in the case studies in this book may be related to industrial output and foreign trade data for the same industrial sectors for which the technology was purchased. It is frequently difficult to match such published economic data and case study material directly, however, and estimates of the impact of such technology transfers may be extremely tenuous in some cases. Nevertheless, this topic still presents itself as a worthwhile area for future research.

BRITISH NATIONAL-ECONOMIC POLICIES

The majority of the trade data relating to British export performance discussed in this book, chiefly refer to the years prior to 1979. Since that year, the financial policies of the British government have led to a comparatively high level of interest rates and a strong pound sterling compared with previous years; and the time horizon of the data presented in this book should consequently be extended as further international trade statistics become available. Such a study should monitor British export performance in general, and to the socialist countries in particular, to investigate the effects of previous and present British national financial policies on her export trade.

RECENT TRENDS IN EAST-WEST POLITICAL AND ECONOMIC RELATIONS

It is apparent that the majority of business arrangements covered in
this study were concluded at a time when East-West inter-governmental
relations were fairly stable and amicable, especially during the period
of détente in the early 1970s. In addition, the socialist countries of
Eastern Europe were regarded as good credit risks during that time, to
whom Western banks were more than willing to lend. These loans were
frequently supported by the respective Western governments to foster a
growth in their countries' exports, and to be seen to be supporting
constructive East-West political and economic relations.

In recent years, however, détente has been seriously damaged as a
consequence of:

(a) the Soviet record on human rights, where several Western governments
    expected rapid improvements following the 1975 Helsinki Conference
    on Security and Co-operation in Europe;
(b) the Soviet intervention in Afghanistan in late 1979;
(c) the establishment of martial law in Poland in late 1981;
(d) the relatively low priority presently given to détente by many
    Western governments.

In addition to these political pressures, future East-West business
contacts will also be under a certain amount of economic strain, caused
by:-

(a) the high costs of Western capital goods to the socialist countries as
    a consequence of continuing Western inflation;
(b) the continuing problems encountered by Eastern European countries in
    marketing their products in the West to raise hard currency;
(c) doubts raised about the creditworthiness of some of the socialist
    countries.

It is this author's view, however, that East-West business relationships
will continue to exist for the main reason described in this book,
namely the company-level responses to the opportunities for further
export trade in products and technology to the socialist countries. The
demand for Western products by the socialist countries may not be as
buoyant in the next few years as in the last two decades, but there is
little evidence to suggest that the Soviet and East European demand for
Western technology has been satisfied. Consequently, demands for
Western know-how for the design and manufacture of end-products and
associated critical components are likely to continue, although purchases
of manufacturing plant from Western countries may be more selective.

A study of the cases contained in this book, together with one of the
author's previous publications (34), suggests that many East-West
business relationships have been built upon contacts extending over a
number of years, which frequently encompass varying political and economic climates. Future East-West business transactions will, therefore,
in this author's view be influenced by the experiences gained during the
periods of rapid trade expansion in the 1960s and 1970s, but directed
towards the needs of East-West commerce in the more restrictive economic
and political climates of the 1980s and 1990s. It is consequently
appropriate to consider and research the forms to be taken by these
future types of East-West business transactions.

NOTES

(1) See Hill (1978), pp.85-135.
(2) See *Business International* (1972), which reviews the experiences of
    a number of Western companies in their business relationships with
    the socialist countries of Eastern Europe; Hannigan and McMillan
    (1979) which reports a survey of Canadian companies engaged in East-
    West trade, based on a postal questionnaire, and Rüthlinshöfer and
    Vogel (1979) which is a study of the Soviet use of machinery imports
    for technology transfer based on the experience of West German ex-
    porters. This study was carried out in parallel with that of the
    present author and P. Hanson, and also supported by Stanford Research
    Institute.
(3) See Hannigan and McMillan (1979).
(4) From Tables 2.2 and 2.10 it can be seen that from 1972 to 1978 the
    UK received 7.5% (i.e. $\frac{10,531}{141,328}$) of the total of exports from the
    centrally planned economies to the developed market economies, but
    exported only 4.9% of the developed market economies deliveries to
    the region over that period. If we take the USSR in particular, we
    can see that the corresponding figures are 9.3% and 3.8% respectively;
    and 6% and 5.8% respectively for the remaining Eastern European
    socialist countries.

    On the other hand, using data available from the 1970/71, 1975 and
    1979 editions of the *UN Yearbook of International Trade Statistics*
    for 1967, 1972 and 1978, it can be seen that although the UK
    accounted for 22% of the value of exports of the SCEEs to the select-
    ed Western nations (i.e. UK, France, Italy, Western Germany, Japan
    and USA) in 1967 this proportion had fallen to 14% in 1972 with no
    change in 1978. Furthermore, although Britain was the largest single
    importer of these selected Western countries, from the socialist
    countries in 1967, it was surpassed by FRG, France and Italy in 1972
    and 1978, with the USA also showing a faster growth rate in imports
    from the socialist countries than the UK from 1967 to 1978; although
    American imports still remained lower than the British.
(5) See, for example, *East-West Markets*, September 19th 1977, p.15
    ('West German Machine Builders Problems'), which discusses some of
    the counterpurchase problems chiefly from the USSR, faced by West
    German machine tool companies.
(6) See, for example, ECE (1979b) and Gutman (1980). Neither of these
    publications on trade and technology transfer refer to any British
    automobile manufacturer as a project leader in an Eastern European
    automobile factory project, with the possible exception of British
    Leyland's activity in Poland.
(7) Gutman (1980) mentions the activities of GKN, Lucas Electrical,
    Girling brakes, and Mintex brake linings.
(8) See data published annually by the United Nations Economic Commiss-
    ion for Europe in *Bulletin of Statistics of World Trade in Enginee-
    ring Products*.
(9) See Hill (1979).
(10) *Ibid.*
(11) Both publications in note (6) above refer to the high degree of in-
    volvement of Fiat and Renault in the USSR (i.e. at Tol'yattigrad,
    the Moskvich car and engine factories, and the Zil and Kamaz truck
    factories).
(12) See *East-West Markets*, September 19th 1977, p.15 ('West German
    Machine Builders Problems'), referred to in note (5) above.

(13) Gutman (1980) refers to the activity of Japanese automotive and
machine tool companies in the socialist countries of Eastern Europe,
which would appear to be consistent with their export drive in world
markets as a whole. Gutman also refers to the setting up of a
special office in USA by Avtopromimport for the specific purpose of
buying American machinery for use in the Kama River Plant.
(14) See McMillan (1977).
(15) See Levcik and Stankovsky (1979), pp.217-222, who base many of their
comments on the results of Bolz and Plotz (1974).
(16) See Paliwoda (1981), pp.167-173.
(17) See Hayden (1976), pp.53-57, and Garland and Marer (1980).
(18) See Hayden (1976), pp.46-53 and ECE (1979a), pp.85-88.
(19) See Macmillan (1971), pp.576,615 and Macmillan (1972), pp.62,64,
for some background information to this agreement.
(20) See Coulbeck (1981) for a discussion of the facilities provided by
ECGD and similar institutions in other Western countries. In
general, the facilities offered by ECGD were found to compare fairly
equally with those of the other Western countries although French
and Japanese government subsidies were higher than those of their
British counterpart for medium term loans.
(21) See footnote (54) to Chapter 2 above for data on British exports of
'machinery non-electric' to the Soviet Union and Eastern Europe,
and Table 8.1 above for information on British exports of machine
tools to the Soviet Union. Similar results are obtained when data
shown in Tables 2.11 and 2.12 above are analysed further.
i.e. British share of exports of engineering goods by the selected
Western nations to:

	1965	1970	1975
(a) USSR	20%*	14%	6%
(b) Remaining SCEEs	16%**	18%	11%

*i.e. $\frac{50}{254}$ x 100%   ** i.e. $\frac{114-50}{652-254}$ x 100%.

(22) For a further discussion of the post 1971 fall in Soviet market
share held by British companies, see Hanson (1980).

(23)
Year	UK Total Exports ($m)	Total Exports of the Developed Market Economics ($m)	UK Total Exports as % of Total Exports
1963	11,790	103,640	11.5%
1970	19,350	224,236	8.6%
1977	54,477	727,709	7.9%

Sources: See Table 2.1 above for 'Total Exports of the Developed
Market Economie s'. UK Total Exports taken from *UN Yearbook of
International Trade Statistics, 1969*, Table B and *UN Yearbook of
International Trade Statistics, 1978*, Table B.

(24)
Year	UK Exports of Engineering Goods ($m)	World Exports of Engineering Goods by Developed Market Economics (DME) ($m)	UK Market Share
1963	5,058	31,040	16%
1970	7,923	78,620	10%
1977	21,542	274,343	8%

Sources: 'UK exports' compiled from *Bulletin of Statistics of
World Trade in Engineering Products* for 1963, 1970, 1977. World
Exports by DMEs from *UN Monthly Bulletin of Statistics*, March 1964,
Special Table C; *1974 Yearbook of International Trade Statistics*,
Table B and *1978 Yearbook of International Trade Statistics*, Table B.

(25)

Year	UK Exports of 'Machinery non-electric' ($m)	Exports of 'Machinery non-electric for Selected Western Nations (UK,France, Italy,FRG,USA & Japan) ($m)	UK Market Share
1963	2,403.8	11,644.1	21%
1970	3,941.1	26,695.8	15%
1977	10,215.1	84,014.9	12%

Source: Compiled from *Bulletin of Statistics of World Trade in Engineering Products*.

(26) From the data shown in Figure 8.1, regression analysis demonstrates that the market share of 'machinery non-electric' exports to the SCEEs of UK, France, Italy, FRG and Japan, compared with the total exports of these products for all five countries can be expressed as:

'Market Share in the SCEEs' = 1.74 + (1.25 x 'Market Share in World Market').

Regression Analysis gives an R^2 Value of 0.83 for this relationship.

(27) For further discussion of the general performance of the British engineering industries, see Saunders (1978) and Report of the Committee of Enquiry into the Engineering Profession (1980).

(28) If USA machinery exports are included with those of the other five Western countries mentioned above, the relationship in note (26) changes to:

'Market Share in the SCEEs' = 10.68 + (0.36 'Market Share in World Market')

For this relationship, the value of R^2 is reduced to 0.10.

(29) For a fuller discussion of this topic see Heiss, Lenz and Brougher, (1979) and US Congress Office of Technology Assessment (1981).

(30) See references quoted in note (29) above.

(31) See Gutman (1981a) and Gutman (1981b).

(32) *Ibid*.

(33) See Zaleski and Wienert (1980), pp.197-240.

(34) See Hill (1978).

Appendices

Appendix A Questionnaire for survey of machine tool imports

QUESTIONNAIRE FOR SURVEY OF SOVIET MACHINE TOOL IMPORTS

Technical background

(a) Could you please describe (in general terms) the main technical features and capacity of the equipment exported to the USSR?

(b) Are any of these technical features covered by patents for which the USSR became a licensee?

(c) If comparable equipment existed already, do you consider that the Soviet engineering industry was capable, at the time when you first discussed this proposal, of producing equipment embodying similar technical features?

If <u>not</u>, could you please give some estimate of the degree of technical lag between Soviet-produced equipment and that designed and manufactured by your company?

If <u>so</u>, why do you consider that the Soviet Union imported equipment from your company?

(d) Can you please give some indication of the degree to which the Soviet customer also had to carry out technical developments to enable your equipment to operate successfully in his manufacturing system?

(e) Was the equipment installed in a new plant or in one already in operation?

Proposal and contract

(a) When was the initial enquiry received?

(b) When was the initial proposal submitted?

(c) Were subsequent proposals necessary?

If <u>so</u>, how did they differ technically from the original proposals?

(d) When was the contract signed?

(e) Was the time taken to reach a finally-agreed proposal longer than, shorter than, or about the same as, the time you would expect to take with, say, a West European customer?

If the time taken was different, to what factor or factors would you attribute the difference?

Acceptance and installation

(a) What were the contractual dates for delivery?

(b) What were the contractual dates for final acceptance on site by the customer, and for meeting the guaranteed standards?

(c) Were these dates observed?

If not, what was the difference in time?

- could you please advance the reasons for this?

- in your experience, were these reasons similar to those encountered in other markets for analogous equipment?

Total time from initial enquiry to installation

(a) What was the difference in time taken from initial enquiry to final acceptance, compared with the time that you would normally expect this entire sequence to take with a West European customer?

(b) If there was a difference, to what factor or factors would you attribute it?

Utilisation

(a) Do you have any information about the utilisation of the equipment after its installation?

If the answer to (a) is Yes:

(b) Was the machinery utilised as fully as you would expect the same machinery to be utilised in a West European factory?

(c) If not, can you advance reasons for this?

(d) If the answer to (b) is Yes, was the time taken to achieve a normal rate of utilisation longer than, shorter than, or about the same as you would expect to observe in a West European company?

(e) If longer, by how much?

(f) If shorter, by how much?

(g) Was the level of manning similar to, greater than, or less than you would expect to observe in a West European company?

(h) Can you comment on any factors that affected the time taken to achieve normal utilisation of the machinery?

(i) Can you comment on any factors that affected manning levels?

Diffusion

(a) Do you know if any basic principles or technical refinements incorporated in the machinery supplied by your company, and not previously utilised in the Soviet Union, have subsequently been incorporated in the design of Soviet-built machinery?

If the answer to (a) is Yes:

(b) Please describe the instances known to you.

(c) What period of time elapsed between the delivery of the British machinery and the production of Soviet machinery incorporating some features of the British machinery?

(d) Would you expect a similar process of learning and 'reproduction' to occur in, say, a West European country?

(e) If the answer to (d) is <u>Yes</u>, would you expect this process to take a longer or shorter time than in the USSR?

(f) Please describe any features of the Soviet industrial scene which you have reason to believe would either hinder or facilitate the spread of successful reproduction of Western technological know-how (e.g. incentive systems, the organisation of research, development and design; and availability of skilled workers and material supplies).

Appendix B Questionnaire for survey of practices in industrial co-operation

QUESTIONNAIRE FOR SURVEY OF PRACTICES IN INDUSTRIAL CO-OPERATION

The number and types of industrial co-operation agreements/arrangements

(a) How many agreements has your company entered into with the socialist countries of Eastern Europe?

(b) Is your Eastern European partner:
    A ministry or state committee?
    A foreign trade organisation?
    A production association or enterprise?

(c) Please indicate the product or activity in which co-operation occurs.

(d) When was the co-operation agreement signed?

(e) In which of the following time ranges does the co-operation agreement fall?
    less than 1 year;   5-10 years;
    1-5 years;   indefinite.

(f) Did the co-operation agreement stem from earlier business with the Eastern European partner?

(g) Which of the following describes the current status of the agreement?
    Under negotiation,
    concluded;   renewed;
    in operation;   replaced by broader agreement;
    expired;   broken off or allowed to lapse.

(h) Name of Western partner(s).

(i) Name of Eastern European partner.

(j) A co-operation agreement may combine a number of activities. Which of the following form part of the agreement in question?
    Provision of managerial services;
    sale of capital equipment;
    sale of complete plant;
    custom design of plant and/or equipment;
    training of Eastern European personnel;
    provision of technical assistance (know-how);
    a licensing agreement;
    supply of parts or components to the Eastern European partner;
    provision by the Eastern European partner of parts or components
      produced to your specification and to be incorporated into your

product;

provision by the Eastern European partner of products produced to your specification and to be marketed by you;

agreement to specialise in the production of certain parts or components and then to exchange them so that each partner can produce the same end product;

agreement to specialise in the production of certain final goods and then to exchange them so that each partner disposes of a full line;

exercise of quality control by you;

an agreement for marketing and servicing in specified geographical areas;

co-operation in a joint project in a third country;

co-ordination of research and development;

exchange of scientific and technical information.

Motivation for Eastern European states to enter into co-operation with British companies

Which of the following motives apply to the Eastern European partner in entering into each agreement?

(a) Expansion, protection or diversification of market opportunities.

(b) Cost savings due to the UK company's advantage in:
    natural resources,
    skilled labour,
    availability of production capacities,
    capital,
    technological expertise,
    managerial expertise,
    other.

(c) Advantages sought in the speedier generation, transfer and application of new technologies including management techniques.

(d) Specialisation and economies of scale.

(e) External economies and 'spin-offs', e.g. the diffusion of modern techniques.

(f) Balanced financing and foreign exchange economies, i.e. by minimising the outlay of foreign exchange the development of trade flows which would probably not otherwise occur is facilitated.

(g) Preferential or privileged treatment of administration, customs or related facilities, e.g. tariffs.

(h) Transport costs, i.e. re-location of production closer to markets or supplies.

(i) Other.

Motivation for British companies to enter into industrial co-operation

Which of the following motives apply to your Company's activity for each arrangement?

(a) Expansion, protection or diversification of market opportunities.

(b) Cost savings due to an Eastern European advantage in:
    natural resources,
    skilled labour,

availability of production capacities,
capital,
technological expertise,
managerial expertise,
other.

(c) Advantages sought in the speedier generation, transfer and application of new technologies including management techniques.

(d) Stability for the formulation and implementation of enterprise plans within a stable and predictable framework, particularly in regard to:
innovation,
investment,
production,
marketing,
other.

(e) Specialisation and economies of scale.

(f) External economies and 'spin-offs', e.g. the diffusion of modern techniques.

(g) Preferential or privileged treatment in respect of administrative, customs or related facilities, e.g. tariffs.

(h) Transport costs, i.e. re-location of production closer to markets or supplies.

(i) Other.

Financial aspects of industrial co-operation

Which of the following elements form part of the financing for each arrangement?

(a) Partial or complete repayment in the goods produced with the equipment and/or technology supplied by you.

(b) Payment arrangements utilising bilateral clearing accounts.

(c) Joint venture financing.

(d) Credit arrangement:
commercial credit arrangements whereby exporter extends credit to importer;
the exporting partner seeks the support of an appropriate banking or financial institution in offering credit to the importer;
the importer seeks the support of an appropriate banking or financial institution in order to pay the exporter.

What difficulties were encountered in making a suitable financial arrangement?

Appendix C Questionnaire for survey of British purchases of Soviet and East European licences

QUESTIONNAIRE FOR SURVEY OF BRITISH PURCHASES OF SOVIET AND EAST EUROPEAN LICENCES

(a) How many licensing agreements has your company entered into with the socialist countries of Eastern Europe?

(b) Is your socialist partner:
   a ministry or state committee?
   a foreign trade organisation?
   a production association or enterprise?

(c) Could you please describe (in general terms) the main technical features of the licence.

(d) When was the agreement signed?

(e) In which of the following time ranges does the agreement fall:
   less than 1 year,   1-5 years,   5-10 years?

(f) Did the agreement stem from earlier business with your Eastern European partner?

(g) Which of the following activities formed part of the agreement in question?
   A licence purchase;
   provision of associated technical assistance (know-how);
   purchase of associated manufacturing equipment;
   purchase of associated components and materials;
   exercise of quality control within your company;
   training of personnel within your company;
   exchange of scientific and technical information;
   an agreement for marketing in specified geographical areas.

(h) Which of the following motives account for your company's interest in the agreement?
   Expansion, protection or diversification of market opportunities through exploitation of the licence;
   cost savings due to an advantage of technological expertise by the socialist partner, in areas relevant to the application of the agreement;
   the speedier application of new technology within the company;
   use of the licence to provide further business contact with the Eastern European partner.

(i) What do you consider to be the main motives for the participation of your Eastern European partner in this business arrangement?

(j) Were the licence practices of the Eastern European partner similar to standard international procedures in terms of:
  method of payment,
  length of agreement,
  guarantees,
  cancellation clauses?

If not, how did they differ?

Appendix D Questionnaire for survey of technical co-operation agreements

QUESTIONNAIRE FOR SURVEY OF TECHNICAL CO-OPERATION AGREEMENTS

(a) How many 'framework agreements' has your company entered into with the socialist countries of Eastern Europe?

(b) Is your socialist partner:
a ministry or state committee,
a foreign trade organisation,
a production association or enterprise?

(c) Could you please describe (in general terms) the main technical aspects covered by the framework agreements?

(d) When was the agreement signed?

(e) In which of the following time ranges does the agreement fall:
less than 1 year,   1-5 years,   5-10 years?

(f) Did the agreement stem from earlier business with the Eastern European partner?

(g) Which of the following activities formed part of the agreement in question?
Exchange of scientific and technical information;
co-ordination of research and development;
training of personnel;
sale of licences;
sale of associated components;
sale of associated manufacturing equipment.

(h) Which of the following motives account for your company's interest in this agreement?
Expansion, protection or diversification of market opportunities through exploitation of the agreement;
cost savings due to certain advantages of technological expertise by the socialist partner, in areas relevant to the operation of the agreement;
the speedier application of new technology within your company;
use of the framework agreement to provide further business contact with the socialist partner.

(i) What do you consider to be the main motives of your socialist partner for participation in this agreement.

Appendix E A case study of West German/Polish industrial co-operation in construction machinery

A CASE STUDY OF WEST GERMAN/POLISH INDUSTRIAL CO-OPERATION IN
CONSTRUCTION MACHINERY

Introduction

The company described in this case study is a West German designer and
manufacturer of construction machinery. Its product range includes
portable and truck-mounted concrete mixers, concrete loaders, concrete
pumps, automation devices, and integrated building systems. The company
has enjoyed particular market success with its range of concrete mixers
accounting for 50 per cent to 60 per cent of the domestic West German
market for portable and truck-mounted concrete mixers, 90 per cent of
West German exports of portable concrete mixers, and 60 per cent of the
West German exports of truck-mounted concrete mixers; although this proportion falls to 5 per cent to 15 per cent for concrete pumps. The
company's truck-mounted mixers have been particularly successful in
Belgium, Scandinavia and Austria, where they have accounted for 75 per
cent to 80 per cent of those countries imports of these product types.
In addition, the company has licensed a Spanish company to manufacture
its truck-mounted concrete mixer under licence, and this product has
secured approximately 85 per cent of the Spanish import market for these
products.

During the mid-1970s, its domestic and export sales market mix changed
quite substantially with a decline in home market sales from 58 million
DM in 1972/73 to 23 million DM in 1974/75. Export sales, on the other
hand, increased from 34 million DM in 1972/73 to 67 million DM in
1974/75. Truck-mounted mixers accounted for approximately 25 per cent of
both domestic and export sales whilst concrete pumps accounted for very
little of domestic sales, and some 8 per cent of export sales.

Domestic sales for the company's range is organised on a product-group
basis, with a head office staff and locally based representatives. The
company exports to over seventy foreign countries, chiefly through
authorised agencies, although the company also has subsidiaries in Paris,
Rotterdam, Beirut (for sales to the Middle East) and Vienna (for sales
to Austria and the socialist countries of Eastern Europe). In addition
to Spain, the company has also licensed the manufacture of its products
to organisations in UK, Yugoslavia, Egypt, South Africa and Kenya. Of
particular interest in this case study is, however, the company's
industrial co-operation agreements with the Polish foreign trade

organisation responsible for the import and export of construction machinery. These agreements relate to the manufacture of the truck-mounted concrete mixer and the concrete pump. The company also has experience in industrial co-operation with Czechoslovakia and Yugoslavia.

Scope of the industrial co-operation agreements

Both of the co-operation agreements included the following common elements:

  training of Polish personnel;
  provision of know-how;
  production of components to the company's specification, by the Polish
    partner;
  exchange of components, in order that each partner may be able to
    assemble a completed product for their own market requirements;
  a marketing agreement covering the rights of each partner in certain
    world regions. The Polish partner had exclusive rights to sell the
    completed products in Poland, and also had marketing rights in the
    CMEA and certain Third World countries, with which the Polish
    government has bilateral trade agreements. The relevant parts of
    these agreements are described in more detail below.

The truck-mounted concrete mixer. The West German company provided know-how, the training of Polish personnel, and the exercise of quality control in Poland to enable their Polish partners to manufacture this product under licence. The company provided the hydraulic pump and the hydraulic motor required for this item, together with some smaller components; and the Polish partner manufactured the mixing drum with its assembly, front and rear supports, and the planetary housing which accounted for approximately 60 per cent by value of the completed product. These components were then exchanged to enable each partner to produce a completed product line.

The concrete pump. As in the agreement described above, the West German company provided the necessary know-how and training to enable the Polish partner to manufacture this item. No licensing fees were required, and the company did not consider it necessary to exercise quality control in Poland in this second agreement, although quality control was exercised in Germany. The company provided all of the necessary hydraulic components for this product, together with some of the components required for its supporting structure, at preferential rates for assembly and re-export. The Polish partner provided other components for the structure, assembled the structure, and delivered the completed frame to the West German company.

Advantages to the Eastern European partner

The main advantages of the industrial co-operation agreement to the Polish partner were that it has been able to secure advanced product development know-how together with completed manufactured products for its domestic and export market requirements, with a lower level of expenditure of hard currency than if the products had been imported. It has also obtained closer links with the Western European partner than in the case of a licensing deal.

Advantages to the Western partner

The West German company obtained the following advantages from its industrial co-operation agreement with its Polish partner:

(a) export of high unit value components to an expanding market;
(b) tariff preferences for components subsequently re-exported in the assembled product;
(c) acquisition of components and assemblies at advantageous prices.

Comments

Although both partners obtained specific advantages from the industrial co-operation agreements, their implementation initially met with certain difficulties, although these were all subsequently resolved. In the first place, problems were encountered in the quality of the product, and there were also delays in delivery. Furthermore, there were sometimes problems encountered in decision-making as a consequence of difficulties in co-ordination between the foreign trade organisation which was the legal partner to the agreement, and the responsible manu-facturing enterprise.

Having solved these initial problems, however, the industrial co-operation agreement was extended in various ways. These have included regular exchange of technical developments, the Polish partner now contributing to technical developments to such an extent that the West German company ceased receiving licensing royalties. In addition, more production was transferred to the Polish partner. Finally, good personal relations at various levels of management were considered to be essential to the success of the agreement, and these had been established to such an extent that the partners were exploring the possibilities for the setting up of joint marketing in some countries, and the establishment of a joint venture in Austria.

Appendix F Market shares for 'machinery non-electric'

MARKET SHARES FOR 'MACHINERY NON-ELECTRIC'

	World Market Share*	SCEE Market Share
UK (1965)	18%	22%
UK (1970)	15%	18%
UK (1975)	13%	8%
France (1965)	7%	13%
France (1970)	8%	14%
France (1975)	10%	16%
Italy (1965)	7%	14%
Italy (1970)	9%	20%
Italy (1975)	8%	14%
FRG (1965)	27%	43%
FRG (1970)	29%	34%
FRG (1975)	29%	40%
USA (1965)	35%	2%
USA (1970)	31%	5%
USA (1975)	30%	13%
Japan (1965)	4%	7%
Japan (1970)	8%	7%
Japan (1975)	10%	9%

*e.g. In 1965 total exports of 'machinery non-electric' by UK, France, Italy, FRG, USA and Japan, was $14309.4 million. World exports by the UK for this product group was $2605 million. Consequently UK market share was 2605/14309% = 18%. Similar calculations were carried out for each country for total world exports, and exports to the SCEEs.

Source: Compiled from data contained in *Bulletin of Statistics of World Trade in Engineering Products*.

Bibliography

Adamson, N., 'The Development of the Design of Large Blast Burnaces', *Steel Times Annual Review*, 1975.

Amann, R., Berry, M.J., Davies, R.W., 'Science and Industry in the USSR', in Zaleski, E., et al. *Science Policy in the USSR*, OECD, Paris, 1969, pp.376-585.

Amann, R., Cooper, J.M., Davies, R.W. (ed), *The Technological Level of Soviet Industry*, Yale University Press, New Haven and London, 1977.

Askansas, B., Fink, G., Levcik, F., *East-West Trade and CMEA Indebtness in the Seventies and Eighties*, Zentralsparkasse und Kommerzbank, Vienna 1979.

Berliner, J., *The Innovation Decision in Soviet Industry*, MIT Press, Cambridge Mass., 1976.

Berry, M.J., Cooper, J.M., 'Machine Tools' in Amann, Cooper and Davies (1977), pp.121-298.

Berry, M.J., Hill, M.R., 'Technological Level and Quality of Machine Tools and Passenger Cars' in Amann, Cooper and Davies (1977), pp.523-563.

Boitsov, V.V., *Mekhanizatsiya i avtomatizatsiya v melkoseriinom proizvodstve*, Mashinostroenie, Moscow, 1972.

Bolz, K., Plötz, P., *Erfahrungen aus der Ost-West-Kooperation*, HWWA-Institut für Wirtschaftsforschung, Hamburg, 1974.

Brown, J.R., 'The Fluid Sand Process: its Place in Modern Foundry Technology', *The British Foundryman*, September 1970, pp.273-279.

Business International, *Doing Business with Eastern Europe*, Business International Publications, Geneva, 1972.

Butslov, M.M., et al. 'Electron-optical method for studying short duration phenomena', *IV^{er} Internationale Kongress Kurzzeitphotographie*, Koln, 1958, pp.230-242.

Bykov, A., 'Perspectives of East-West Relations in Technology Transfer and Related Problems of Dependence', in Saunders (1977), pp.165-178.

Chase World Information Corporation, Series on East-West Business Co-operation and Joint Ventures, *Hungary*, New York, 1976.

Chase World Information Corporation (CWIC), *Chase World Information Series on East-West Business Co-operation and Joint Ventures: USSR*, CWIC, New York, 1977.

Chernykh, V.P., *Vliyanie spetsializatsii na uroven' proizvoditel' nosti truda*, Ekonomika, Moscow, 1965.

Clarke, D., 'Pioneer in World Aviation', *Touchdown 3*, BAC, Weybridge.

Coulbeck, N., *The UK Export Credit System in Relation to Trade with the Socialist Countries of Eastern Europe : A Comparative Study*, unpublished paper, Department of Management Studies, Loughborough

University of Technology, July 1981.

Council for Mutual Economic Assistance (CMEA), *Statistical Yearbook of Member States of the Council for Mutual Economic Assistance*, Statistika, Moscow (English language edition published by IPC, London, 1978 and 1979.

Dashchenko, A.I., Nakhapetyan, E.G., *Proektirovanie, raschet i issledovanie osnovnykh uzlov avtomaticheskikh linii i agregatnykh stankov*, Nauka, Moscow, 1964.

Davies, R.W., 'The Technological Level of Soviet Industry : an overview' in Amann, Cooper and Davies (1977) p.66.

Day, A.J., *Exporting for Profit*, Graham and Trotman, London, 1976.

Demchenko, M.N., *Spetsializatsiya i proizvoditel'nost truda*, Nauka, Moscow, 1964.

Dragilev, M.S. et al., *Problemy razvitiya ekonomicheskikh otnoshenii mezhdu sotsialisticheskimi i kapitalisticheskimi stranami*, Moskovskii gosudarstvennyi universitet, Moscow, 1974.

Economic Commission for Europe (ECE), *Analytical Report on Industrial Co-operation*, ECE, Geneva, 1973.

Economic Commission for Europe (ECE), Committee on the Development of Trade, *Legal Forms of Industrial Co-operation Practised by Countries Having Different Economic and Social Systems with Particular Reference to Joint Ventures*, (Trade/AC.3/R.10), ECE, Geneva, 1976 (ECE (1976a).

Economic Commission for Europe (ECE), Committee on the Development of Trade, *A Statistical Outline of Recent Trends in Industrial Co-operation*, Trade/AC.3/R8, ECE, Geneva, 1976 (ECE 1976b).

Economic Commission for Europe (ECE), Committee on the Development of Trade, *Industrial Co-operation and Transfer of Technology between ECE Member Countries* (Trade/AC.3/R9), ECE, Geneva, 1976 (ECE 1976c).

Economic Commission for Europe (ECE), Committee on the Development of Trade, *Promotion of Trade through Industrial Co-operation; Statistical Outline of Recent Trends in Industrial Co-operation*, (Trade/R.373/Add. 5) ECE, Geneva, 1978 (ECE 1978a).

Economic Commission for Europe (ECE), Committee on the Development of Trade, *Promotion of Trade through Industrial Co-operation: The Experience of Selected Western Enterprises Engaging in East-West Industrial Co-operation, Results of a Survey of Fifteen Firms in the Machine Tool Sector*, (Trade/R.373/Add.4), ECE, Geneva, 1978 (ECE 1978b).

Economic Commission for Europe (ECE), *East-West Industrial Co-operation*, (ECE/Trade/132), United Nations, New York, 1979 (ECE 1979a).

Economic Commission for Europe, *East-West Co-operation in the Automotive Sector and Counter-trade Arrangements*, Trade/R.385/Add.1), ECE, Geneva, 8th October, 1979 (ECE 1979b).

Gardner, C., *British Aircraft Corporation - A History*, B.T. Basford, London, 1981.

Garland, J., Marer, P., 'US Multinations in Poland : A Case Study of the International Harvester - BUMAR Co-operation in Construction Machinery' in US Congress Joint Economic Committee, *East European Economic Assessment, Part 1 - Country Studies 1980*, US Government Printing Office, Washington, 1981, pp.121-137.

Gutman, P., 'Co-operation industrielle Est-Ouest dans l'automobile et modalités d'insertion des pays de l'est dans la division internationale du travail occidentale' 1st and 2nd parts published in *Revue d'études comparatives est-ouest*, Vol. 11, No. 2 (June 1980), pp.99-154 and Vol. 11, No. 3 (September 1980), pp.57-100.

Gutman, P., 'Tripartite Industrial Co-operation and Third Countries' in Saunders, C.T. (ed), *East-West-South : Economic Interactions between*

Three Worlds, Macmillan, London, pp.337-364 (Gutman 1981a).

Gutman, P., 'Tripartite Industrial Co-operation and East Europe' in US Congress Joint Economic Committee, *East European Economic Assessment Part 2 - Regional Assessments*, US Government Printing Office, Washington, 1981, pp.823-871 (Gutman 1981b).

Hannigan, J.B., McMillan, C.H., *The Participation of Canadian Firms in East-West Trade : A Statistical Profile*, Research Report No. 11, East-West Commercial Relations Series, Institute of Soviet and East European Studies, Carleton University, Ottawa, 1979.

Hanson, P., 'International Technology Transfer from the West to the USSR' in US Congress Joint Economic Committee (1976), pp.786-813.

Hanson, P., 'The Trade Partner Composition of Soviet Trade with the West' in Fallenbuchl, Z., McMillan, C.H., *Partners in East-West Economic Relations*, Pergamon, New York, 1980, pp.185-219.

Hanson, P., *Trade and Technology in Soviet-Western Relations*, Macmillan, London, 1981.

Hanson, P., Hill, M.R., 'Soviet Assimilation of Western Technology : A Survey of UK Exporters' Experience' in US Congress Joint Economic Committee (1979), pp.582-604.

Hayden, E.W., *Technology Transfer to Eastern Europe : US Corporate Experience*, Praeger, New York, 1976.

Hayward, G., Lethbridge, D., *European Case Studies in Business Policy - A Workbook*, Harper Row, London, 1975.

Heiss, H.W., Lenz, A.J., Brougher, J., 'United States-Soviet Commercial Relations since 1972' in US Congress Joint Economic Committee (1979), pp.37-53.

Hill, M.R., *Standardisation Policy and Practice in the Soviet Machine Tool Industry*, unpublished PhD thesis, University of Birmingham 1970.

Hill, M.R., 'The Administration of Standardisation in the USSR, *The Quality Engineer (Journal of the Institute of Quality Assurance)*, Vol. 37, No. 2, pp.39-43 (Feb. 1973) (Hill 1973a).

Hill, M.R., 'Standardisation in the Soviet Machine Tool Industry', *Production Technology Abstracts and Reports from Eastern Europe*, No. 45, pp.3-9 (Dec. 1973) (Hill, 1973b).

Hill, M.R., *The Export Marketing of Capital Goods to the Socialist Countries of Eastern Europe*, Wilton Publications (Gower), Farnborough 1978.

Hill, M.R., 'Desk Research for the Soviet Capital Goods Market', *European Journal of Marketing*, Vol. 13, No. 8 (1979), pp.271-283.

Hill, M.R., 'International Industrial Marketing into Eastern Europe', *European Journal of Marketing*, Vol. 14, No. 6, 1980, pp.139-164.

Holzman, F.D., 'Foreign Trade Behaviour of the Centrally Planned Economies' in Roskovsky, H. (ed), *Industrialisation in Two Systems : Essays in Honour of Alexander Gerschenkron*, New Yori, 1966, pp.237-263.

Hovyani, G., Ellis, P.S., 'Parallel Views regarding the Case of an East-West Industrial Co-operation', *Papers of the 1978 Congress of the European Society of Opinion and Marketing Research*, ESOMAR, Amsterdam, 1978, pp.535-564.

Hutchings, R., *Soviet Science, Technology, Design; Interaction and Convergence*, RIIA/Oxford UP, London, 1976.

Institution of Production Engineers (I. Prod. E.), *Improvements Required in Machine Tools*, I. Prod. E., London, 1968.

Kemenes, E., *Revue de l'Est*, 1974, No. 2, pp.89-105, 138-139.

Kiser, J.W., *Report of the Potential for Technology Transfer from the Soviet Union to the United States*, Department of State and National Science Foundation, Washington, 1977.

Kiser, J.W., *Commercial Technology Transfer from Eastern Europe to the*

United States and Western Europe, US Department of State/Kiser Research, Washington DC, 1980.

Kiser, J.W., 'Tapping Eastern bloc technology', *Harvard Business Review*. March-April 1982, pp.85-93.

Knirtsch, P., 'Vom Ost-West Handel zur Wirtschaftskooperation', *Europa-Archiv*, 1973, No. 3.

Langrish, J. et al., *Wealth from Knowledge*, Macmillan, London, 1972.

Levcik, F. (ed), *International Economics - Comparisons and Interdependencies*, Springer-Verlag, Vienna, New York, 1978.

Levcik, F., Stankovsky, J., *Industrial Co-operation between East and West*, M.E. Sharpe Inc., White Plains, New York, 1979.

Liass, A.M., *Foundry Trade Journal*, Vol. 124, No. 3 (1968).

Macmillan, H., *Riding the Storm, 1956-1959*, Macmillan, London, 1971.

Macmillan, H., *Pointing the Way, 1959-1961*, Macmillan, London, 1972.

McMillan, C.H., 'East-West Industrial Co-operation' in US Congress Joint Economic Committee, *East-European Economies post-Helsinki*, US Government Printing Office, Washington, 1977, pp.1175-1224 (McMillan 1977a).

McMillan, C.H., 'Forms and Dimensions of East-West Inter-firm Co-operation' in Saunders (1977), pp.28-60 (McMillan 1977b).

McMillan, C.H., 'Growth of External Investments by the COMECON Countries' *The World Economy*, Vol. 2, No. 3, (September 1979), pp.363-386, (McMillan 1979a).

McMillan, C.H., 'Direct Soviet and East European Investment in the Industrialised Western Economies' in US Congress Joint Economic Committee (1979), pp.625-647 (McMillan 1979b).

Olszynski, J., 'Kooperacja przemyscowa Polski z Wysoko rozwinietymi Krajami Europy Zachodniej', *Zeszyty naukowe*, 1973, No. 94 (Warsaw SCPi S).

Paliwoda, S., *Joint East-West Marketing and Production Ventures*, Gower, Aldershot, 1981.

Pinder, J., Pinder, P., *The European Community's Policy Towards Eastern Europe*, RIIA/PEP, London, 1975.

Portes, R., Winter, D., Burkett, J., *Macroeconomic Adjustment and Foreign trade of Centrally Planned Economies*, Paper presented at the Econometric Society World Congress, Aix-en-Provence, August 1980.

Radice, H., 'Experiences of East-West Industrial Co-operation : A Case Study of UK firms in the Electronics, Telecommunications and Precision Engineering Industries' in Saunders (1977), pp.144-151.

Report of the Committee of Enquiry into the Engineering Profession (chaired by Sir M. Finniston), *Engineering our Future*, HMSO, London 1980 (Cmnd. 7794).

Richman, B., 'Multinational Corporations and the Communist Nations', *Management International Review*, 1976, No. 3, pp.9-22.

Rogers, L.J., 'Practical problems in dealing with Eastern European nations', *Procurement Weekly*, Vol. 8, No. 3, (16th January, 1980), pp.8, 9.

Röthlingshöfer, K.Ch., Vogel, H., *Soviet Absorption of Western Technology : Report on the Experience of West German Exporters*, Report submitted to Stanford Research Institute, Menlo Park, California, 1979.

Saunders, C.T. (ed), *East-West Co-operation in Business : Inter-firm Studies*, Springer-Verlag, Vienna and New York, 1977.

Saunders, C.T., *Engineering in Britain, West Germany and France*, University of Sussex European Research Centre, 1978.

Scott, N., 'The Scope for Industrial and Technological Co-operation on a Large Scale', in Watts, N.G.M. (ed), *Economic Relations between East and West*, Macmillan, London, 1978, pp.222-230.

Starr, R. (ed), *East-West Business Transactions*, Praeger, London, 1974.
Sternheimer, S., *East-West Technology Transfer : Japan and the Communist Bloc*, The Washington Papers, Vol. 8, No. 76, Sage Publications, 1980.
Sutton, A.C., *Western Technology and Soviet Economic Development, 1945-1965*, Hoover Institution Press, Stanford, California, 1973.
Szita, J., *Perspectives for All-European Economic Co-operation*, A.W. Sijthoff, Leyden, 1977.
Tabaczynski, E., *Revue de l'Est*, 1974, No. 3, pp.27-35.
Theriot, L.H., 'US Governmental and Private Industry Co-operation with the Soviet Union in the Fields of Science and Technology' in US Congress Joint Economic Committee (1978), pp.739-766.
Turpin, W.N., *Soviet Foreign Trade*, D.C. Heath, Lexington, Mass., 1977.
US Congress Joint Economic Committee, *Soviet Economy in a New Perspective* US Government Printing Office, Washington, DC, 1976.
US Congress Joint Economic Committee, *Soviet Economy in a Time of Change*, Vol. 2, US Government Printing Office, Washington, DC, 1979.
US Congress Office of Technology Assessment, *Technology and East-West Trade*, Allanheld Osmun, Montclair N.J./Gower, Farnborough, 1981.
Volchkevich, L.I., *Nadezhnost' avtomaticheskikh linii*, Mashinostroenie, Moscow, 1969.
Wasowski, S., *East-West Trade and the Technology Gap*, Praeger, New York, 1973.
Wilczynski, J., *The Economics and Politics of East-West Trade*, Macmillan, London, 1969.
Wilczynski, J., *Technology in Comecon*, Macmillan, London, 1974.
Wilczynski, J., *Licences in the West-East-West Transfer of Technology*, unpublished paper, University of New South Wales Faculty of Military Studies, Department of Economics, 1976 (Wilczynski, 1976a).
Wilczynski, J., *The Multinationals and East-West Relations*, Macmillan, 1976, (Wilczynski, 1976b).
Wilczynski, J., 'Licences in the West-East-West Transfer of Technology', *Journal of World Trade Law*, Vol. 11, No. 2 (March/April 1977), pp.121-136.
Yarkov, A.M. (ed), *Naladka i ekspluatizatsiya avtomaticheskikh linii iz normalizovannykh uslov*, Mashinostroenie, Moscow, 1965.
Yergin, A.S., *East-West Technology Transfer : European Perspectives*, The Washington Papers, Vol. 8, No. 75, Sage Publications, 1980.
Young, J., *Quantification of Western Exports of High Technology Products to Communist Countries*, Industry and Trade Administration, Office of East-West Policy and Planning, US Department of Commerce, Project No. D-41.
Zaleski, E., Wienert, H., *Technology Transfer between East and West*, OECD, Paris, 1980.

Index

References from Notes indicated by 'n' after page reference.

Adamson, N. 147n
Aerospace and Aircraft
  Industries 78, 83-85, 95-97,
  102-120, 144-146
Agricultural Engineering 78,
  83-85, 94, 95, 100n, 120-126,
  131n, 169
Amann, R. 2, 3, 11n, 72n
Askansas, B. 44n, 45n
Austria 39
Austrian companies 174-176
Automotive engineering 3,
  49-71, 78, 83-85, 87-90, 101n,
  162, 163, 172-174, 177, 178

BAC 1-11 103-120
Berliner, J. 11n, 12n
Berry, M.J. 72n, 73n
Boitsov, V.V. 72n
Bolz, K. 193n
Brainard, L.J. 11n
British Aerospace plc. 102-120,
  162, 165, 180
British Cast Iron Research
  Association (BCIRA) 143
British Shipbuilders 129, 131n,
  132n
Britten-Norman Ltd. 104-106,
  108-110
Brougher, J. 194n
Brown, J.R. 147n
Bulgaria (including Bulgarian
  foreign trade organisations)
  7-10, 40, 41, 87-89, 100n,
  123, 134, 136-139, 156, 169
Burkett, J. 2, 11n
Business International S.A. 159n
Butslov, M.M. 159n
Bykov, A. 37, 45n

Carleton University, Ottawa 37-39,
  77, 166, 167
Chase World Information Corporation
  (CWIC) 45n, 73n
Chemical industry 39, 99n
Chemicals (and petrochemicals) 3,
  20, 22, 25-27, 29, 33, 34, 36,
  39, 149
Chernykh, V.P. 72n
Clarke, D. 130n
Coal industry 81-83, 99n, 100n,
  153-155
Coalmining machinery 78, 81-83,
  153-155
Computers 3, 78, 90-93, 174
Confederation of British Industry
  (CBI) 10
Co-ordinating Committee of NATO
  (COCOM) 11n, 152
Construction engineering machinery
  169, 170, 206-208
Cooper, J.M. 72n
Coulbeck, N. 193n
Council for Mutual Economic
  Assistance (CMEA or COMECON) 3,
  9, 37, 44n, 68, 93, 95, 99n, 123,
  142, 152
Counterpurchase 75-124, 162-179
Cranes 78-81, 99n, 169
Czechoslovakia (including
  Czechoslovakian foreign trade
  organisations) 2, 7, 8, 10, 36,
  85, 87-89, 100n, 103, 134, 150,
  153, 169

Dashchenko, A.I. 72n
Davies, R.W. 2, 3, 11n, 72n
Day, A.J. 11n
Demchenko, M.N. 72n

Department of Industry 80
Department of Trade 10, 133
Dragilev, M.S. 77, 99n

Economic Commission for Europe
  (ECE) 5, 6, 12n, 37-39, 45n,
  48n, 78, 99, 166, 167, 193n
Electrical engineering 36, 39,
  40, 78, 85-87, 99n, 100n,
  141-143, 150, 151, 171, 172
Electronics 3, 36, 39, 40, 134,
  135, 150-152, 156, 157
Ellis, P.S. 99n
Emgert, M. 11n
Engineering products (imports
  and exports) 20, 21, 27, 28,
  33-35, 43, 180-189
European Economic Community
  (EEC) 9
Export Credit Guarantee
  Department (ECGD) 96, 97, 111,
  121, 123, 126, 179

Federal Republic of Germany (FRG
  or Western Germany) and West
  German companies 38, 39, 41,
  44, 46n, 59, 63-65, 110, 111,
  139, 149, 163, 164, 169-172,
  179-189
Fink, G. 44n, 45n
Fork-lift trucks 87-89
France (including French
  companies) 39, 41, 46n, 96,
  139, 149, 163, 164, 173, 174

Gardner, C. 130n
Garland, J. 193n
German Democratic Republic
  (GDR or Eastern Germany) 2,
  7, 8, 36, 40, 41, 94, 134,
  156, 169
GKN Contractors Ltd. 124-126,
  165
Gutman, P. 131n, 187, 190,
  192n-194n

Hannigan, J.B. 161, 192n, 193n
Hanson, P. 11n, 45n, 73n, 77
Hayden, E.W. 11n, 193n
Hayward, G. 130n
Heiss, H.W. 194n
Hill, M.R. 11n, 12n, 45n, 73n,
  76, 158n, 159n, 192n, 194n
Holzman, F.D. 11n
Hovyani, G. 77, 99n
Hungary (including Hungarian
  foreign trade organisations)

2, 7, 8, 36-39, 79, 81-87, 93-95,
  99n, 100n, 134, 141-143, 150, 153,
  169
Hutchings, R. 11n

'Islander' aircraft 104-106, 108-110
Industrial co-operation 5-7, 37-39,
  75-132, 163-179
Institution of Production Engineers
  (I. Prod. E.) 73n
Iron and steel industries 3, 135-141,
  143-146
Italy (including Italian companies)
  39, 41, 44, 47n, 149, 163, 164,
  179-189

Japan (including Japanese companies)
  39, 41, 44, 47n, 87, 140, 149,
  163, 164, 179-189, 193n
Judy, R.W. 11n

Kemenes, E. 45n
Kiser, J. 36, 45n, 133, 134, 139,
  144, 146, 147n
Knirtsch, P. 45n

Langrish, J. 74n
Lenz, A.J. 194n
Lethbridge, D.G. 130
Levcik, F. 23, 38, 44n, 45n, 76, 77,
  99n, 169, 193n
Liass, A.M. 147n
Licences 33, 36, 78-81, 83-97,
  163-179

Machine tools 3, 4, 12n, 39, 40,
  49-71, 99n, 105, 107, 124, 129-131n,
  163, 164, 170
Macmillan, H. 12n, 193n
McMillan, C.H. 6, 39, 40, 45n, 75,
  77, 78, 99n, 133, 141, 147n, 149,
  159, 161, 166, 167, 192n, 193n
Marer, P. 44n, 193n
Massey-Ferguson-Perkins Ltd. 120-124,
  162, 163, 165, 180
Mechanical engineering 36, 39, 44,
  99n, 180-189

Nakhapetyan, E.G. 72n
National Coal Board (NCB) 81, 153-
  155, 158
Nuclear engineering 3, 41, 78, 93,
  94

Oil industry 155, 156
Olszynski, J. 38, 45n

Paliwoda, S.J. 130n, 169, 193n
Pinder, J. and P. 12n
Plötz, P. 193n
Poland (including Polish foreign
  trade organisations) 2, 7, 8,
  10, 36-39, 78-81, 83-85, 89-93,
  99n-103, 120-130, 134, 153,
  156, 162, 169, 180
Portes, R. 2, 11n

Radice, H. 101n
Rechargeable batteries 141-143
Reich, M. 11n
Report of the Committee of Enquiry
  into the Engineering Profession
  (Finniston Report) 194n
Richman, B. 75, 99n
Rogers, L.J. 130n
Romania (including Romanian
  foreign trade organisations)
  7, 8, 10, 38-41, 81, 83, 89, 90,
  93, 94, 101n-120, 126-130, 134,
  156, 162, 169, 180
Rötlingshofer, K.Ch. 192n

Saunders, C.T. 194n
Scott, N. 45n
Stankovsky, J. 23, 38, 45n, 76,
  77, 99n, 169, 193n
Starr, R. 6, 12n
Sutton, A.C. 3, 11n
Sweden 39, 44, 138, 139, 176-178
Szita, J. 45n

Tabaczynski, E. 45n
Technical co-operation 7, 37-40,
  87-89, 148-159, 190
Theriot, L.H. 46n, 148, 158n, 159
Turpin, W.N. 11n

UK Atomic Energy Authority (UKAEA)
  93
UK Inter-governmental Agreements
  with the SCEEs 40, 41, 158,
  179, 180
US Office of Technology Assessment
  45n, 194n
USA (including American companies)
  39, 41, 44, 47n, 86, 96, 108,
  109, 139, 149-152, 164, 179-191
USSR (including Soviet foreign trade
  organisations) 2-5, 8-10, 15-20,
  27, 33-41, 49-71, 81, 84, 88,
  90, 93, 134-136, 139-158, 162,
  163, 179, 180

Vogel, H. 192n
Volchkevich, L.I. 72n

Wasowski, S. 3, 11n, 45n
Wienert, H. 3, 11n, 45n, 194n
Wilczynski, J. 3, 11n, 33, 37,
  44n, 45n, 75-77, 99n, 133, 136,
  147n
Winter, D. 2, 11n

Yarkov, A.M. 72n
Yergin, A.S. 12n
Young, J. 45n
Yugoslavia 33, 36, 37, 78, 79, 96

Zaleski, E. 3, 11n, 45n, 158n, 194n